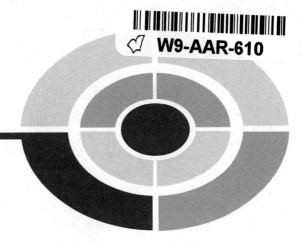

Anatomy Demystified

Demystified Series

ANATOMY DEMYSTIFIED

**The Hon. Dr. Dale Pierre Layman, Ph.D.,
Grand Ph.D. in Medicine (Belgium)**

McGRAW-HILL

New York   Chicago   San Francisco   Lisbon   London
Madrid   Mexico City   Milan   New Delhi   San Juan
Seoul   Singapore   Sydney   Toronto

The McGraw·Hill Companies

Cataloging-in-Publication Data is on file with the Library of Congress

1 2 3 4 5 6 7 8 9 0   DOC/DOC   0 9 8 7 6 5 4

ISBN 0-07-143827-0

The sponsoring editor for this book was Judy Bass and the production supervisor was Pamela Pelton. The art director for the cover was Margaret Webster-Shapiro. It was set in Times Roman by Keyword Publishing Services Ltd.

Printed and bound by RR Donnelley.

 This book was printed on recycled, acid-free paper containing a minimum of 50% recycled, de-inked fiber.

McGraw-Hill books are available at special quantity discounts to use as premiums and sales promotions, or for use in corporate training programs. For more information, please write to the Director of Special Sales, Professional Publishing, McGraw-Hill, Two Penn Plaza, New York, NY 10121-2298. Or contact your local bookstore.

This book is fondly dedicated to my wonderful wife, Kathy. She is the "Anchor," the solid "Rock" upon which we can all lean in times of trouble or need. She is my best and most patient friend.

I also wish to highlight Allison, one of my artistically talented daughters. It is through her creative efforts that our host – Professor Joe, The Talking Skeleton – has been brought to vibrant life.

Finally, I wish to thank Janet M. Evans, President of the American Biographical Institute, and Nicholas S. Law, Director General of the International Biographical Centre (Cambridge, England). In many of their volumes, they have described my ideas for "Intuitive Geometry and the A&P (Human Anatomy & Physiology) Text."

And, now, through the help of Judy Bass and Scott Grillo at McGraw–Hill, these ideas are being realized!

CONTENTS

CONTENTS

CONTENTS

CONTENTS

PREFACE

"Which comes first—the chicken, or the egg?" This book is about the "egg"—*human anatomy* or body *structure*. (An unhatched egg is an example of anatomy.) Its close companion, *PHYSIOLOGY DEMYSTIFIED*, is all about the "chicken"—*human physiology* or body *function*. (The process of laying an egg is an aspect of chicken physiology, after all!)

ANATOMY DEMYSTIFIED is for people who want to get acquainted with the fundamental concepts of human body structure, without having to take a formal course. But it can also serve as a supplemental text in a classroom, tutored, or home-schooling environment. In addition, it should be useful for career changers who need to refresh their knowledge of the subject. I recommend that you start at the beginning of this book and work straight through.

This book seeks to provide you with an intuitive, highly visual grasp of anatomy and its terminology. It takes the approach used by many of the early visual artists. It travels far, far back in time, back to the Time of the Ancient Pharaohs, and of their Great Pyramids rising majestically from the sands of the Egyptian desert! It is here that much of the formal study of anatomy first began, so it is here that *we* begin!

The book takes you on a *Journey Through Bodyspace*, represented by the Great Body Pyramid! Professor Joe, The Talking Skeleton, is our host. He is drawn as a cartoon standing upright and pointing, whenever key facts about Biological *Order* in the human body are being presented. But when he is fallen and fractured, our Good Professor is talking to you about facts of Biological *Disorder* in the human body. These key facts of Order-versus-Disorder will be about Anatomy, Physiology, or just Plain Body Functions.

As you go from body system to body system, you will also learn *where* to put many facts of Biological Order/Disorder, briefly writing them within the "grids or drawers" of the Great Body Pyramid. In this way, like putting your socks into a drawer, you will always know where to find the key body facts ("socks") whenever you need them!

This introductory work also contains an abundance of practice quiz, test, and exam questions. They are all multiple-choice, and are similar to the sorts of questions used in standardized tests. There is a short quiz at the end of every chapter. The quizzes are "open-book." You may (and should) refer to the chapter texts when taking them. When you think you're ready, take the quiz, write down your answers, and then give your list of answers to a friend. Have the friend tell you your score, but not which questions you got wrong. The answers are listed in the back of the book. Stick with the chapter until you get most of the answers correct.

This book is divided into six sections. At the end of each section is a multiple choice test. Take these tests when you're done with the respective sections and have taken all the chapter quizzes. The section tests are "closed-book," but the questions are not as difficult as those in the quizzes. A satisfactory score is three-quarters of the answers correct. Again, answers are in the back of the book.

There is a final exam at the end of this course. It contains questions drawn uniformly from all the chapters in the book. Take it when you have finished all six sections, all six section tests, and all of the chapter quizzes. A satisfactory score is at least 75 percent correct answers.

With the section tests and the final exam, as with the quizzes, have a friend tell you your score without letting you know which questions you missed. That way, you will not subconsciously memorize the answers. You can check to see where your knowledge is strong, and where it is not.

I recommend that you complete one chapter a week. An hour or two daily ought to be enough time for this. When you're done with the course, you can use this book, with its comprehensive index, as a permanent reference. What you now hold in your hand, I think you will agree, is a most *unusual* approach to the study of human anatomy!

We have of course, our most unique talking skeleton host, Professor Joe (and occasional glimpses of his somewhat mischievous sidekick, Baby Heinie). More importantly, this book represents the practical application of what I like to call *Compu-think*, or "*compu*ter-like modes or ways of human *think*ing." This is reflected in its heavy emphasis upon binary (two-way) classifications, grid-associated reasoning, and frequent occurrence of summary word equations.

Suggestions for future editions are welcome.

Now, work hard! But, be sure to have *fun*! Best wishes for your success.

The Hon. Dr. Dale Pierre Layman, PhD, Grand PhD in Medicine

ACKNOWLEDGMENTS

The most interesting and entertaining illustrations in this book are mainly due to the talented efforts of one of my own daughters, Allison Victoria Layman. It is through her gifted eyes that my visions for picturing key body concepts have been successfully brought to life!

I extend thanks to Emma Previato of Boston University, who helped with the technical editing of the manuscript of this book. Thanks also go to Maureen Allen and the staff at Keyword, who helped me winnow out various errors and inconsistencies within the original manuscript.

I particularly wish to thank Judy Bass, Senior Acquisitions Editor at McGraw-Hill. Judy has been *very* enthusiastic and supportive of all my writing efforts! This means *a lot* to a struggling author! She also deserves credit for her brilliant insight that we need two separate but closely-linked books—both an *ANATOMY DEMYSTIFIED*, as well as a stand-alone *PHYSIOLOGY DEMYSTIFIED*—in this series.

Finally, Mr. Scott Grillo (publisher) has been a quiet, steadfast, and kindly presence behind all of my writing efforts. I am most pleased to honor both Judy Bass and Scott Grillo within the distinguished pages of the *Dedication Section* in *2000 OUTSTANDING INTELLECTUALS OF THE 21ST CENTURY* (2nd Ed., 2003), just published by the International Biographical Centre (Cambridge, England). The unusual and creative thinking efforts being presented in these humble little volumes are closely being watched and reported on, within the Highest World Intellectual Circles!

ABOUT THE AUTHOR

Dale Layman, Ph.D., Grand Ph.D. in Medicine, is a popular professor of biology and human anatomy and physiology at Joliet Junior College in Illinois. He is the author of many articles, as well as *PHYSIOLOGY DEMYSTIFIED* and *BIOLOGY DEMYSTIFIED*, two closely related books in this series. He is the first Grand Doctor of Philosophy in Medicine for the United States, and has received many other international honors and awards for his writing, teaching, and highly creative thinking. Dr. Layman has more than 28 years of experience in the biological sciences.

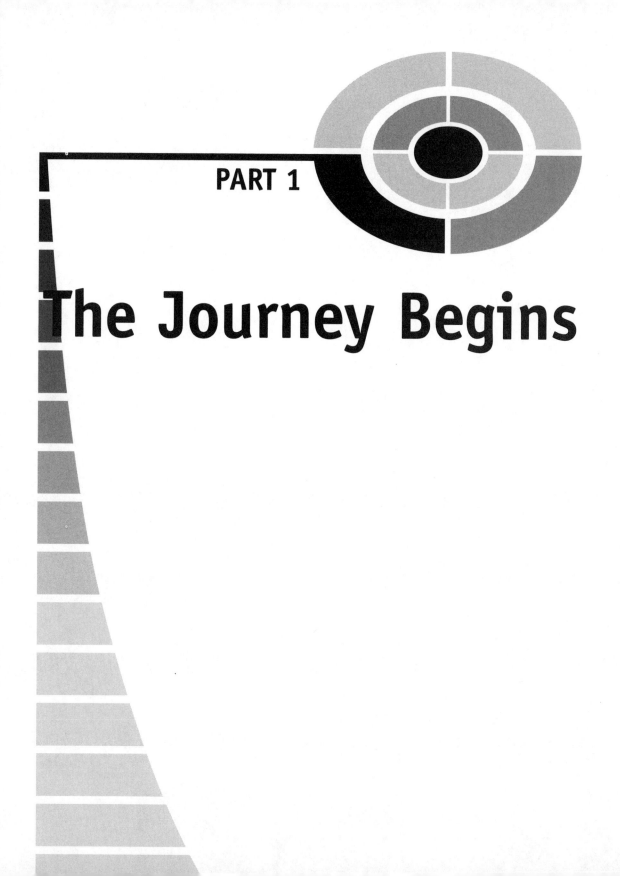

PART 1

The Journey Begins

Anatomy: Our Inner World of Bodyspace

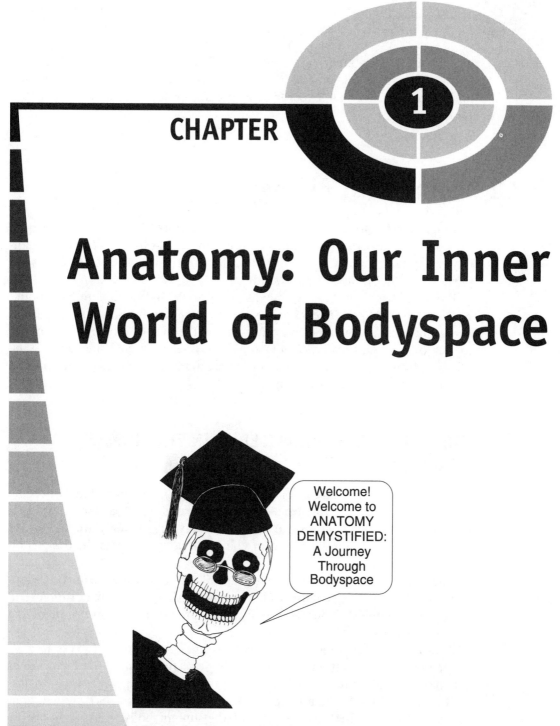

Welcome! Welcome to ANATOMY DEMYSTIFIED: A Journey Through Bodyspace

Hello, there! I am Professor Joe, The Talking Skeleton! I have been selected as your guide for this book, *ANATOMY DEMYSTIFIED*. I am here to give you a basic, "bare bones" introduction to The Place Below Your Skin! You and I are about to take a wondrous, almost magical trip through the human body. Together, we shall experience a special journey ... *A Journey Through Bodyspace*.

It All Begins with Biology

Before we get into *anatomy* (pronounced as **ah**-**NAT**-oh-me), we need to introduce the broader subject of *biology* (buy-**AHL**-oh-jee). The word, biology, is actually a technical term that comes from two Ancient Greek word parts – *bi* ("life") plus -*ology* ("study of").

Biology, therefore, literally means the "study of life." (A related book, *BIOLOGY DEMYSTIFIED*, covers this topic in detail.) Since our subject is *human* anatomy, we will be concentrating on the part of biology that applies to our own species, called *Homo sapiens* (**HOH**-moh **SAY**-pea-**ahns**). The phrase, Homo sapiens, derives from Ancient Latin. It exactly translates to mean, "man or human being having wisdom."

BIOLOGICAL ORDER: PATTERNS IN THE HUMAN ORGANISM

As soon as you start studying biology (or anatomy), you immediately become aware of its many *patterns*. In general, a pattern is some particular arrangement of shapes, forms, colors, or designs. [**Study suggestion:** Carefully examine some of the patterns you find in your own home, such as the designs on your wallpaper.]

The Homo sapiens species consists of all us human *organisms* (**OR**-gan-**izms**). An organism in general is a living body with a high degree of *Biological Order*. By Biological Order, we simply mean a recognizable pattern involving one or more organisms.

[**Study suggestion**: Find a photo, or just imagine, a flock of geese in a V-shaped pattern of flight. Is this pattern a case of Biological Order? How does this type of order differ from that found in your wallpaper?]

Remember that our main topic is the human organism. So, we will concentrate upon the Biological Order of the human body: that is, upon the particular patterns found within a single human organism. (Examine the specific body patterns pictured in Figure 1.1.)

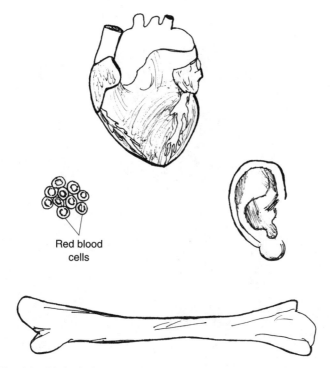

Red blood cells

Fig. 1.1   Biological Order: Some patterns within the human body.

BIOLOGICAL ORDER —— A PARTICULAR PATTERN WITHIN THE HUMAN BODY

    You should be able to easily name and identify the various body structures whose form or pattern has been drawn in Figure 1.1. In this book, key chapter facts about Biological Order in the human body will be tagged in the page margins by a miniature icon version of me, Professor Joe (Figure 1.2), looking "orderly" with my fine, straight pointer.

BIOLOGICAL DISORDER: A BREAK IN HUMAN BODY PATTERNS

Just as we can classify certain facts as being examples of Biological *Order*, we can also classify exactly *opposite* facts as instances of Biological *Dis*order!

Fig. 1.2   Professor Joe standing upright: An icon for biological order.

Consider, for instance, the long bone shown back in Figure 1.1. Its pattern is whole and intact. Thus, the bone is easily recognizable as a particular case of Biological Order in the body.

But what happens when such a long bone is hit extremely hard, and it fractures? Just as an intact bone pattern represents Biological Order, a fractured bone represents a particular case of Biological Disorder. In general, the following word-equation applies:

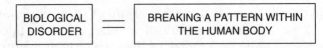

| BIOLOGICAL DISORDER | — | BREAKING A PATTERN WITHIN THE HUMAN BODY |

In this book, key facts about Biological Disorder in the human body will be tagged in the page margins by a miniature icon version of me, Professor Joe – fractured and beheaded – with my straight pointer now sadly broken! (see Figure 1.3.)

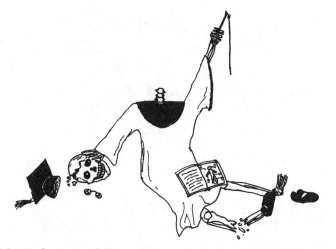

Fig. 1.3   Professor Joe fallen and fractured: An icon for biological disorder.

The Process of Dissection: Anatomy is a Real "Cut-up"!

So far, we have outlined the importance of asking a single question about key facts in this book: "Is this particular body fact an example of Biological Order, or is it an example of Biological Disorder?"

An important follow-up question to ask is, "Does the fact also represent anatomy, or does it represent *physiology* (**fih**-zee-**AHL**-uh-jee)?" To answer this second question, we have to look far, far back in history. Anatomy probably began in the days of Ancient Greece and Rome. The word, anatomy, exactly translates from Ancient Greek to mean, "the process of" (-*y*) "cutting" (*tom*) something "up or apart" (*ana*-).

A term closely related to anatomy is the Latin word, *dissection* (dih-**SEK**-shun). The prefix, *dis*-, means "apart" (like the *ana*- in anatomy). The word root or main idea, *sect*, translates to mean "cutting" (identical to the *tom* in anatomy). Finally, the suffix or word ending, -*ion*, means "the process of" (just like the suffix, -*y*, in anatomy). One basic relationship thus becomes immediately apparent:

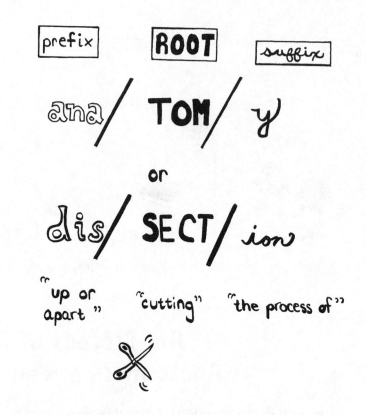

VESALIUS AND HIS CADAVERS

We can clearly see that, speaking from history, anatomy seems to be all about dissection. Yet, in the time of Ancient Greece and Rome, dissections of human beings were rarely, if ever, done. (Once in a while, a dissection was carried out on the body of a gladiator slain in the arena.) Early physicians were reluctant to dissect, because the human body was considered to be the sacred vessel of the spirit or soul. Hence, what the early Greek and Roman philosophers knew about anatomy was mainly derived from dissection of apes and other animals!

One of the first public dissections of an actual *human* body took place in the year 1341, at the medical school in Padua (**PAD**-you-ah), Italy. Over the next 200 years, there was a rather frequent, but not very thorough, dissection of human *cadavers* (kuh-**DAV**-ers) – bodies that have "fallen dead" (*cadav*).

One who pioneered the thorough dissection of human cadavers was an Italian professor named *Andreas* (an-**DRAY**-us) *Vesalius* (vih-**SAY**-lee-**us**). Living from 1514 to 1564, Vesalius started out by dissecting the bodies of executed criminals. Quite a few bodies came straight from hanging on the gallows – right to Vesalius' dissection table! Here, by the process of patient and painstaking dissection, the "cut-up" bodies of the criminal cadavers finally revealed many long-hidden secrets of human anatomy to Vesalius and his curious students (see Figure 1.4).

Fig. 1.4   Vesalius reveals human anatomy by dissecting cadavers.

BODY STRUCTURES BECOME ANATOMY

Although Vesalius did not actually begin the dissection of human cadavers, he is frequently given credit as the *Father of Anatomy*. An important reason was that he wrote the first scientific textbook on human anatomy, called *On The Fabric of The Human Body,* or *Seven Books on the Structure of the Human Body*. This pioneering work contained detailed drawings of actual body structures, which were sketched by artists while Vesalius, himself, performed careful dissections. For thousands of years up to this time, scholars seldom performed adequate human dissections, preferring to read the flawed works of the Ancient Greeks and Romans, instead. These earlier works were mainly based upon dissections of animals (not humans), and they included much error-filled speculation.

From the title of his book, you will note that Vesalius made a rough equality among three different things:

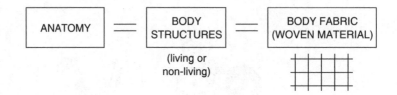

Vesalius equated anatomy with *body structures*. Thus, we can simply define anatomy as body structure and the study of body structures. But to understand this definition deeply, we need to be quite clear about a *structure*. Looking carefully at Figure 1.5, you will find 7 common types of structures. [**Study suggestion:** Before reading any further, try to identify and name each of the seven types of structures shown.]

The seven structures are properly named in left-to-right sequence as follows: a woven, checkered cloth fabric; human red blood cells; globe of the Earth; a deer skull; a human skull; a black rubber boot; and a section of human skin.

Fig. 1.5   Seven common types of structures.

Basic characteristics of structures

Structures such as these have five basic characteristics:

1. A structure takes up space. Whenever a structure is present, it takes up all three dimensions of space. That is, it has some amount of length, width, and height. And since they *do* occupy space, there cannot be an infinite number of structures (such as black rubber boots) present within a particular limited amount of space (such as a hallway closet).

2. A structure has mass and weight. A structure like a black rubber boot contains *mass* – a certain amount of "matter" within its tall sides, toebox, and flat-soled bottom. And in places with gravity (such as the planet Earth), a structure also has weight. (Picture yourself struggling to run a 26.2-mile marathon race wearing a pair of heavy, clunking, black rubber boots!)

3. A structure is literally "built-up" from a smaller number of parts. The word, structure, actually means a "building" up of something from a number of smaller parts. The human skull is a structure, for example, because it is "built-up" from a considerable number of smaller bone parts.

4. A structure assumes a particular size, shape, and color. Besides consisting of smaller parts, a structure often has these parts colored in a certain shading. And these colored parts are then added together and rearranged to produce a single larger structure having a certain shape and size. A black-and-white checkered tablecloth, for instance, consists of dozens of square units – some colored black, others white. When these little squares are sewn together in straight rows and straight columns, such that their combined length is significantly greater than their width, a rectangle-shaped tablecloth results.

5. The underlying "skeleton" of a structure can be modeled as a woven cloth fabric or a grid – one consisting of intersecting rows and columns of square units. As you may recall from the title of his famous book, Vesalius closely tied anatomy and body structure with a "woven fabric" of material. Reviewing Figure 1.5, note that the black-and-white checkered tablecloth actually consists of many interwoven horizontal rows and vertical columns of square units, when it is magnified and viewed close-up. Similarly, the deep *connective tissue layer* of the human skin consists of many *connective tissue fibers*. These thin, rod-like structures interlace with one another like a criss-crossing fabric of interwoven cloth.

But even when the close-up, magnified view of a particular structure does *not* show such a *real* "woven-fabric" pattern, the structure can still be *modeled* as if it had this woven pattern. Consider, for example, the common practice of dividing the surface of a plastic globe of the Earth into sets of intersecting grid lines having particular latitudes and longitudes. This widely-accepted technique of geometric modeling into some number of square or rectangular units is a very convenient way of "dissecting" a particular structure for careful study. One obvious advantage of this approach is that you can always know "where" you are in the total fabric or grid skeleton of the structure, at any particular time.

LIVING BODY FUNCTIONS BECOME PHYSIOLOGY

If you could classify a structure (or body structure) as part of a sentence, what particular part would you choose? Would the structure be a noun, or would it be a verb? Consider this simple sentence: "The hammer hit the nail." Both the hammer and nail are structures, and they both serve as nouns in the sentence.

Now, what about the word, *hit*? This word is an action verb, isn't it? It is also classified as a *function*. A function, in general, can be defined as something that a particular structure does, or something that is done to the structure. In our sample sentence, the word, *hit*, is something that the hammer (a structure) does. But in terms of the nail, *hit* is something that is done to it. In either case, the action verb, *hit*, is considered a type of function. This idea of action is closely related to the literal translation of the Latin word, function, which means "perform" (*doing* something).

"What about physiology?" you might now ask. "How does physiology differ from function, in general?" The word, physiology, actually translates from Latin to mean, "the study of" (*-ology*) "Nature" (*physi*). More broadly, the term also means, "natural science." This reflects the ancient idea that physiology was the study of practically *everything* that was in Nature, including such non-living things as rocks and stars! But as the centuries passed, human knowledge accumulated to the point where physiology became restricted to the study of only the *living* things that were in Nature – such as human beings, plants, fish, and animals.

Therefore, physiology can be briefly defined as the study of *living* body functions: that is, the study of the Nature of *living* things.

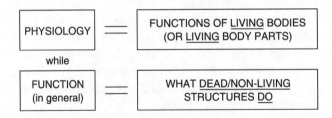

A sample sentence for physiology might be, "The little boy hit the nail with his hammer." As before, the nail and hammer are both structures (sentence nouns). The verb, *hit*, however, is here an example of physiology (body function) rather than just plain function (performance or action verb). This is because the little boy represents anatomy (body structure) that is living and carrying out a body function (hitting the nail with a hammer). (A companion volume to this book, *PHYSIOLOGY DEMYSTIFIED*, describes the World of Physiology in much greater detail.) [**Study suggestion:** Does a cadaver lying on a dissection table still have anatomy? Does it still have physiology? – Why, or why not? (Review the definitions of anatomy versus physiology to help you, if necessary.).]

A Two-way System of Classification for Body Facts

Summarizing our progress so far, we have now created a two-way system for classifying key body facts:

1. Biological Order versus Biological Disorder. Either a key body fact in this book is an example of Biological Order (recognizable pattern), or it is an example of Biological Disorder (broken pattern). An intact Professor Joe, The Talking Skeleton, is used in the margins to tag recognizable patterns. And a fractured, fallen-down Professor Joe is employed to tag broken body patterns.

2. Body Structure (Anatomy) versus Living Body Function (Physiology). In addition to each key fact representing either Biological Order or Disorder, it also represents either Body Structure (Anatomy) or Living Body Function (Physiology).

Since the title of this book is *ANATOMY DEMYSTIFIED*, we certainly want to teach ourselves how to clearly distinguish facts of body *structure* from those of body *function*, don't we! If we can't successfully make this fundamental distinction, then we are still mostly "mystified" about the subject matter! To help us in this important thinking task while we read, we will have our handy Professor Joe icon appear with a second icon – either a black capital "*A*" for an *Anatomy* fact, or a white capital "*P*" for a *Physiology* fact. (In certain cases where the particular body structures being discussed are *not* living, we will substitute a white letter "*F*" for just plain body *Function*, instead of Physiology.)

A

anatomy

P

physiology

F

function

The Different *Types* of Anatomy

We have been discussing body structure or anatomy on the one hand, contrasted with living body functions or physiology, on the other. Now it is time to delve into the fascinating Domain of Anatomy in considerably greater depth. A key concept here is that anatomy always involves the study of body structures. But the particular *type* of body structures being studied may sometimes differ a lot!

COMPARATIVE ANATOMY = BODY STRUCTURES OF *ALL* TYPES OF ORGANISMS

Since human bodies were considered sacred in many early societies, the first dissections were mainly carried out on other types of creatures. We have already mentioned, for example, that monkeys and apes were often "cut apart" by the Ancient Greeks and Romans to study their *internal* ("inside") pattern of body parts. Therefore, a *comparative* (come-**PAIR**-uh-tiv) *anatomy* has existed since Ancient Times. The word, comparative, actually means "pertaining to" (*-ive*) "comparing" (*comparat*). In comparative anatomy, then, the body structures of different kinds of animals are carefully *compared* with one another. But to be very technical, comparative anatomy not only involves *animal* body structure but also the body structure of *plants*. Hence, we can state that:

COMPARATIVE ANATOMY = COMPARING BODY STRUCTURES
OF *ALL* TYPES OF ORGANISMS
(but primarily those of different animals)

Comparative anatomy focuses upon discovering both the similarities and the differences that exist between the body structures of various animals. In general, the more closely related two different types of animals are, the more closely their overall shape tends to follow a *common body plan*. Consider, for instance, the skulls of two *mammals* (**MAM**-als) shown among the seven types of structures back in Figure 1.5. Both the human skull and the deer skull are quite similar, except, of course, for the presence of antlers! One major reason for this fairly close resemblance is the common body plan of the mammals – backboned organisms with hair and "breasts" (*mamm*).

The grid or matrix background skeleton: a "womb" or "mother" for comparing body structures

The search for common body plans among closely related groups of organisms is very much helped by the use of "skeletal" grids or "woven fabrics" as backgrounds behind the shapes of body structures. A grid is also called a *matrix* (**MAY**-tricks), which is Latin for "womb" or "mother." A well-known anatomist named *D'Arcy Thompson*, for instance, in 1961 published a famous book entitled, *On Growth and Form*. D'Arcy Thompson was one of those who thoroughly re-adopted the old idea of "woven fabrics" from Vesalius and used rectangle-shaped grids or *matrices* (**MAY**-truh-**sees**) as background overlays upon pictures of the body structures of humans and animals. Thompson employed intersecting x (horizontal) and y (vertical) lines to create a rectangular matrix with numbered square cells. When overlaid upon pictures of, say, chimp and human skulls (Figure 1.6), such a matrix could be deformed or twisted slightly so that the anatomy of one type of organism could be directly compared with the anatomy of another. Thompson repeatedly demonstrated this general principle: the more closely related two types of organisms are, the less deforming or twisting of their overlaying grid skeleton is required to transform their skulls or other body structures into one another. Thus, we can conclude that deer are less closely related to humans than are chimps, because the grid pattern of the human skull is more easily deformed to match a chimp skull than it would have to be to match a deer skull.

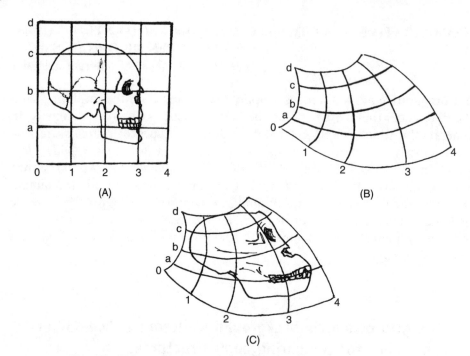

Fig. 1.6 Using grids to compare chimp and human skull anatomy.

EMBRYOLOGY – TALKING ABOUT SOME "SWELLING" FELLERS!

Comparative anatomy, as we have seen, concentrates upon the similarities and differences among the shapes of various organisms. *Embryology* (**em-bree-AHL-uh-jee**), alternately called *Developmental Anatomy*, can also look at the differences between diverse creatures. Here, however, the emphasis is upon the growth and development of *embryos* (**EM-bree-ohs**).

An embryo is literally "a sweller." It is a tiny, early stage of development that progressively grows or "swells" into a newborn human, plant, or other animal. Embryology ("the study of swellers") is very useful in comparative anatomy, especially in regards to *evolution*. The term, evolution, means "a process of rolling out." Studies in comparative anatomy repeatedly demonstrate that the early embryos of various organisms – such as humans, reptiles, and amphibians – share some elements of a common body plan. Humans and turtles, for example, both have early embryos with tails! This suggests that, far, far back in time, both humans and turtles are related to a common ancestor (or group of ancestors). Or, if this is not the case, then embryology strongly hints that both humans and turtles had similar environmental forces acting upon them, so the anatomy of their embryos *adapted* to these forces in a similar way. (A companion volume, *BIOLOGY DEMYSTIFIED*, discusses the Theory of Evolution, adaptation, and the fossil record, in much more detail.)

HUMAN ANATOMY: *OUR* INNER WORLD OF "BODYSPACE"

This book emphasizes *normal* anatomy in the *human adult*, rather than comparative anatomy or embryology. And since the body structures of human anatomy all occupy space, we are going to frame the subject as if we were dividing up some particular area of space. In fact, we are going to introduce the human organism as our Inner World of "*Bodyspace*."

We are going to take a gridded or matrix-like approach to dividing up Bodyspace. Such an approach is taken by a farmer who plows up the earth in his fields to create rectangular rows and furrows of crops. And it is used by the modern city planner, who subdivides a town by overlaying a grid of intersecting streets upon it. We are essentially making use of *geometry* (jee-**AHM**-uh-tree) – "the process of" (-*y*) "earth measurement" (*geometr*).

GALILEO, GEOMETRY, AND THE INTERNAL ENVIRONMENT

Speaking of space, how about outer space? It is here, in the vast universe lying far beyond the "earth" (*geo*), that the planets and other "heavenly bodies" twirl around. After all, it was *Galileo* (**gal**-uh-**LEE**-oh), the famed Italian astronomer and observer of moons and planets, who said that, "The Book of Nature is written in characters of Geometry." Since Galileo also studied medicine, he obviously would apply geometry to the study of our *internal environment* as well – the "inner space" of the entire body lying deep to the surface of the skin. It is this internal environment, or "inner space" of the human body, which we have called Bodyspace. This volume of Bodyspace is important, because all of the internal structures of the body are found within it.

ANCIENT EGYPTIANS AND ITALIAN ARTISTS "DISSECT" BODYSPACE INTO GRIDS

Galileo, who lived from 1564 to 1642, was certainly not the first person in the world to see "The Book of Nature" (especially the internal environment of the human body) as "written in characters of Geometry." The early roots of geometry lie far back in time, even before the Dawn of Recorded History!

How do we know this? We only have to look at the walls of ancient tombs, which housed the mummified bodies of the Ancient Egyptians. Preserved for centuries in the dry air of the desert, here one can still admire the elegant square grids on the inner tomb walls, carefully marked out by using strings dipped in red paint (Figure 1.7, A). Special talented workers, called outline scribes, then used black paint to carefully sketch pictures of humans, beasts, and gods onto the gridded wall. The underlying skeletal grid or matrix pattern provided a highly ordered framework that helped the scribes intuitively "see" how to draw the shapes of the various anatomical objects in their proper proportions.

Thousands of years later, extending even into today, visual artists have often employed rectangular "sighting-grids" to help them "see" the human body in proper perspective for accurate drawings. One such visual artist was named *Albrecht Dürer*, who adopted the *velo* ("veil") technique of the Italian artists to frame the nude body within a sighting grid (Figure 1.7, B).

In this book, *ANATOMY DEMYSTIFIED*, we, too, will often use a grid to help us efficiently organize information and to "dissect" the human body by use of geometry – why, we're going to teach ourselves this body information and think like the Ancient Egyptians! This simple procedure will allow us to "see" more intuitively and to thoroughly understand the key facts of human body structure.

HUMAN BODYSPACE: A SERIES OF STACKED GRIDS IN SPACE

Figure 1.7 presented *superficial* (soo-per-**FISH**-al) or "pertaining to the surface" diagrams of the human form, as drawn by visual artists – both ancient and modern. These diagrams, however, just reflected the *external* ("outside") appearance of the body *surface*. Now, what about the internal environment, that part of the human body lying deep *within*? If the surface of the human form can be drawn upon the orderly background of a grid with square cells, then cannot *Human Bodyspace* – the internal environment – be modeled as a series of horizontal grids? These horizontal grids could be conveniently stacked, one upon the other.

(A)

(B)

Fig. 1.7   The use of grids for drawing the human body: (A) Ancient Egyptians and (B) visual artists.

The Different *Levels* of Anatomy

Overall, if we are going to model Human Bodyspace (the internal environment) as a stacked series of horizontal grids, we need to know what geometric form the entire stack will create. For guidance, let us turn back, way back in time.

THE ANCIENT EGYPTIANS GIVE US THEIR GREAT PYRAMIDS

We find ourselves in Ancient Egypt, back in the time of the Great Pyramids.

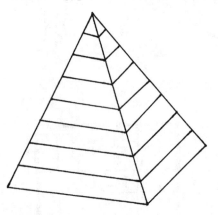

The pyramid shape consists of a number of horizontal levels, each level stacked upon a lower one. Therefore, here is our ideal model for Human Bodyspace!

THE LEVELS OF BODY ORGANIZATION = STACKED GRIDS IN A GREAT PYRAMID

Let us use this form of the Ancient Egyptians and have it symbolize the various *levels of body organization*. A level of body organization represents a certain degree of size and complexity of body structures. When stacked together as horizontal slabs, these individual levels of body organization can be seen as creating a Great Body Pyramid. If such a Body Pyramid has nine (IX) different levels, then the lower levels represent smaller and simpler body structures, while the higher levels denote larger and more complex body structures (Figure 1.8, A). Further, if each level or slab is pulled out of the Great Pyramid, we see that it consists of a grid or matrix of square cells (Figure 1.8, B).

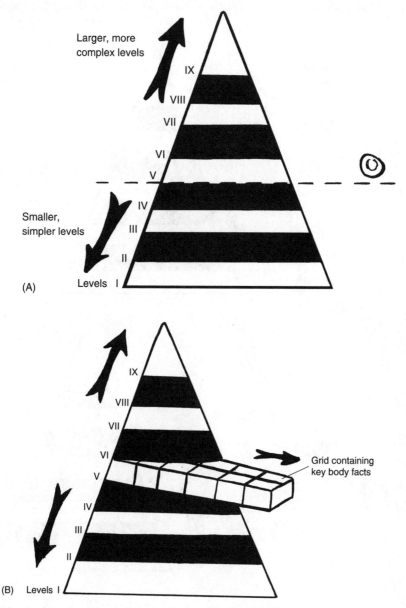

Fig. 1.8 An introduction to the great body pyramid. (A) The great body pyramid. (B) Each level of body organization: square cells within a grid.

For our purposes, however, we actually want to distinguish between two different Body Pyramids. *The Great Pyramid of Structural <u>Order</u>* is contrasted with *The Great Pyramid of Structural <u>Disorder</u>*.

The Great Pyramid of Structural Order (Figure 1.9, A) includes all nine levels of structural *organization* found within the human body. Conversely,

Fig. 1.9   The two great body pyramids: One of order, one of disorder. (A) The great pyramid of structural order. (B) The great pyramid of structural disorder.

The Great Pyramid of Structural *Dis*order includes the same nine levels, but each now represents structural *disorganization* that can sometimes be found within the human body (study Figure 1.9, B).

In summary,

THE GREAT PYRAMID OF STRUCTURAL *ORDER:*
"Normal" **Body Patterns**
versus
THE GREAT PYRAMID OF STRUCTURAL *DISORDER:*
"Broken" **Body Patterns**

It is the intact Professor Joe who is standing in the margins of the book pages, pointing to key text facts that belong to particular levels or grids in The Great Pyramid of Structural Order. And it is a fallen-down Professor Joe, who is all beaten-up and fractured, who points to key text facts belonging to certain levels in The Great Pyramid of Structural Disorder. For convenience, it is good to think of each key fact as occupying a certain square cell within a horizontal grid, the grid being a particular level of body organization. (Picture each level of body organization as a drawer being pulled out of a Great Pyramid, and the drawer in turn being subdivided into many square cells or chambers.)

The nine levels

Forming the broad base of the Great Pyramid are the *subatomic* (sub-ah-**TAH**-mik) *particles* (see Figure 1.10). This broad base is followed in sequence by a stack of eight higher levels of body organization, each of them representing structures of progressively greater size and complexity. Specifically, these levels are the *atoms*, *molecules*, *organelles* (**OR**-gah-**nels**), *cells*, *tissues*, *organs*, *organ systems*, and finally, the entire *human organism*.

Carefully observe from Figure 1.10 (B) that the *Cell Level* (*V*) is given a special name – "The Life-line." This is because the cell represents the lowest *living* level of body organization.

GROSS (MACROSCOPIC) ANATOMY = BODY STRUCTURES WE CAN *SEE*

Your *gross* (**GROHS**) income (before taxes) is always "bigger" than your net income (after taxes), isn't it? Thus, gross means "big" or "large." A synonym

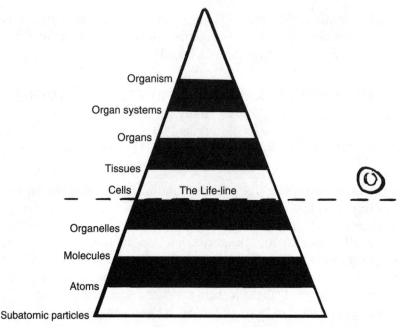

Fig. 1.10   The nine levels of body organization.

(similar word) for gross is *macroscopic* (**MA**-kroh-**skahp**-ik). This word literally "refers to" (*-ic*) structures that appear "large" (*macr*) when they are "examined" (*scop*). Since they are so large, then, the body structures studied in *gross* or *macroscopic anatomy* are all clearly visible to the naked eye. In the human body, gross (macroscopic) anatomy generally refers to Levels VI through IX – the Tissue Level up through the entire Organism Level (Figure 1.11, A). We humans can clearly see, for example, the reddish-brown *cardiac* (**KAR**-dee-**ak**) *muscle tissue* present within the "heart" (*cardi*) wall, as well as the whole organ of the heart, itself. Thus, both the cardiac muscle tissue and the heart organ are studied in gross (macroscopic) anatomy.

MICROSCOPIC ANATOMY = BODY STRUCTURES WE *CANNOT* SEE

If you look at the heart organ, and dig around in the reddish-brown-colored cardiac muscle tissue, you will still not be able to directly view the individual cardiac muscle *fibers* – actually thin, fiber-shaped, *cardiac muscle cells*. This is because in the human body, the Cell Level (V) is part of *microscopic* (**migh-kroh-SKAHP**-ik) *anatomy*.

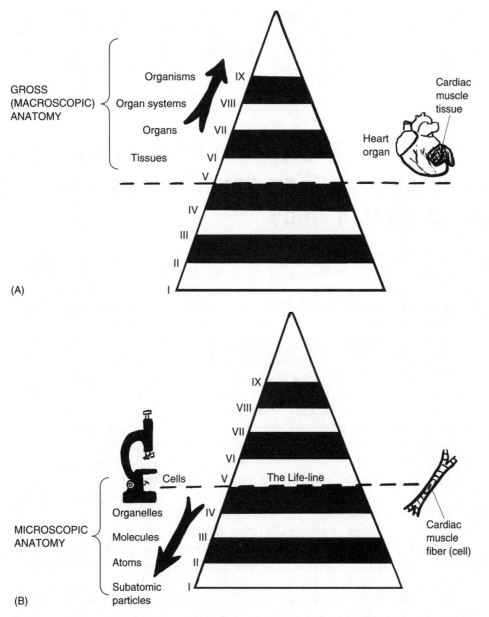

(A)

(B)

Fig. 1.11   The various levels of anatomy: Gross (macroscopic) versus microscopic. (A) Levels of gross (macroscopic) anatomy within the great body pyramid. (B) Levels of microscopic anatomy within the great body pyramid.

Microscopic body structures are those that appear "tiny" (*micr*) when they are "examined" (*scop*). In fact, the body structures classified as microscopic anatomy include all those levels from the cell on down (Figure 1.11, B). These tiny body structures are not visible to the naked human eye, so they must be carefully viewed under microscopes of various kinds. [**Study suggestion:** Go ahead and try to list the particular levels of body organization that are classified as microscopic anatomy. When done, check these levels with those shown in Figure 1.11, B.]

Summarizing Our Basic Task: Distinguishing *A* from *P*, Order from Disorder, Level-by-Level, Within the Human Body

In this first chapter of the book, we have been steadily building an organized system for helping you to really learn and "demystify" the subject of anatomy. An important starting part of this system, you may recall, is the use of the Professor Joe icon to help us distinguish between facts representing Biological Order, and those representing Biological Disorder.

Following closely on this contrast between Order (intact body patterns) and Disorder (broken or disrupted body patterns), we pointed out that it was also essential to distinguish exactly *what* in the body was associated with these patterns. Specifically, we proposed using a capital letter *A* in the page margins to denote key facts about *anatomy*. The letter *P* tags key facts about *physiology*, and the letter *F* indicates key facts about just plain *functions* (those not involving the functions of *living* body structures).

The combined result, therefore, is a *two-way system of fact classification*. We have classification by Order versus Disorder, *and* classification by body structure (anatomy) versus body function (physiology or just plain function). For key facts about anatomy involving Biological Order, then, we will have Professor Joe standing upright, with a capital A under his pointing stick (Figure 1.12, A). For facts about anatomy that are in a state of Biological Disorder, in contrast, the fallen Professor Joe will have a capital A under his broken pointer (Figure 1.12, B).

Taking a similar approach, key facts about physiology in a state of Biological Order will be tagged with standing Professor Joe having a white capital P under his pointer (Figure 1.12, C). And facts about physiology being in a condition of Disorder will be identified by fallen Professor Joe with a

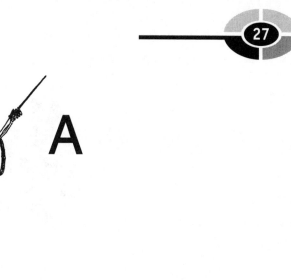

(A)

(B)

Fig. 1.12 Is Professor Joe standing or fallen, and is it his anatomy, or just his function? (A) Biological order, anatomy. (B) Biological disorder, anatomy. (C) Biological order, physiology. (D) Biological disorder, physiology. (E) Biological order, plain function. (F) Biological disorder, plain function.

white P under his broken pointer (Figure 1.12, D). Key facts about just plain function will be treated the same way as those for physiology. But they will be tagged with a capital F rather than P under standing Professor Joe's pointer, or under fallen Professor Joe's broken pointer (Figure 1.12, E and F).

(C)

(D)

Fig. 1.12   (continued)

(E)

(F)

Fig. 1.12   (continued)

BODY LEVELS IN THE GREAT PYRAMID: ANATOMY (AND FUNCTION) STARTS MUCH "LOWER" THAN PHYSIOLOGY!

You may now be asking yourself, "Okay, I see how this two-way system of fact classification works, but *how* am I supposed to know when a key fact represents *physiology*, rather than just plain *function*?" The answer, once again, can be found by returning to the Great Pyramid, and closely examining its horizontal levels of body organization.

Figure 1.13 does this nicely, for us. Observe that Level V, the Cell Level, is where physiology really starts. "Why is this?" the curious reader may wonder. *Physiology doesn't start until the Cell Level is reached, because the cell is the lowest **living** level of body organization. And physiology involves the function of **living** body structures, only.* (A companion volume to this book, *PHYSIOLOGY DEMYSTIFIED*, provides thorough descriptions of the characteristics of living body functions. Another companion, *BIOLOGY DEMYSTIFIED*, defines "living" in general.)

"I understand," you retort, now becoming more enlightened. "But where does just plain *function* start?" *Plain function starts at the same place that*

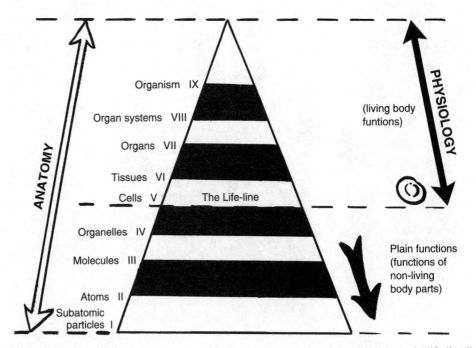

Fig. 1.13 Anatomy and plain function begin before the cell and its physiological "life-line."

anatomy does – at the very lowest level of body organization. From Figure 1.13, it is clear that the subatomic particles within the human body are the lowest examples of anatomy. If we describe what they do, or what happens to these subatomic particles, then plain functions are involved.

Such plain functions also occur at the successively higher levels of atoms, molecules, and organelles, because these levels (like the subatomic level) involve *non-*living body structures. *Therefore, anatomy (and associated plain functions) begin at a much lower (**non-living**) level on the Great Body Pyramid, compared to physiology and its activities of **living** body structures.*

The final result of all this knowledge is our ability to create a *three-way system of fact classification.* In addition to Professor Joe and the letters *A*, *P*, or *F* under his pointer, we will write in the appropriate level of body organization involved.

Consider, for instance, a normal long bone (Figure 1.14, A). Key facts about this normal bone would be tagged by standing Professor Joe (Biological Order), with the letter *A* (anatomy) and the word, *Organ*, placed below his pointer. This is because a bone represents the organ level of body organization. (This level will be defined in a later chapter.) Let us consider a specific example of such a key fact: "Long bones are usually found in the body limbs or extremities."

Organ 1

In contrast, think about a fractured long bone (Figure 1.14, B). Key facts about this fractured bone would be tagged by a fallen, broken Professor Joe (Biological Disorder). If the fact describes something happening to the long bone when it is fractured, then it is tagged with the letter *P* (physiology) and, once again, *Organ*, placed below his broken pointer. For instance, consider this key fact: "A fractured long bone may be dangerous, especially if it breaks through the overlying skin." Here we have a case of physiology, since this sentence includes an action verb that describes something happening – the fractured bone *breaking* through the overlying skin.

Organ 1

In conclusion, our three-way system of classification helps us to effectively categorize or pigeonhole key body facts according to their state of Biological Order or Disorder; their existence as either anatomy, physiology, or plain function; and their particular level of body organization. Helpful *Body-Level Fact Grids* are found at the end of each chapter. Each of these Body-Level Grids is like an open drawer or matrix of square cells within the Great Body Pyramid. Each grid provides a specific place for you to briefly write-in and summarize the key facts you have read, about a particular level of body structure. As you progress through this book, chapter-by-chapter, you can conveniently return to these Body-Level Grids and review and retrieve the information stored there. In this way, you will be learning to think much like an Ancient Egyptian, who liked to put almost everything somewhere into a Pyramid!

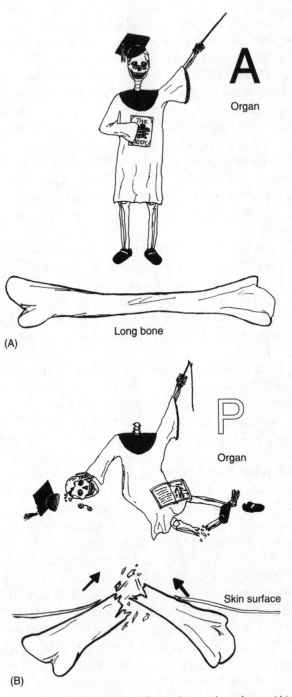

Fig. 1.14   A three-way system for classifying facts about a long bone. (A) Biological order, anatomy, organ level. (B) Biological disorder, physiology, organ level.

Quiz

Refer to the text in this chapter if necessary. A good score is at least 8 correct answers out of these 10 questions. The answers are listed in the back of this book.

1. The word, *biology,* literally translates from Latin to mean:
    (a) "Breaking of bones"
    (b) "Study of life"
    (c) "Study of animals and plants"
    (d) "Body structures and functions"

2. An organism can be correctly defined as:
    (a) A small, dead creature without significant functions
    (b) "Man or human being having no wisdom"
    (c) The lowest living level of human body organization
    (d) A living body with a high degree of Biological Order

3. Good examples of body patterns include:
    (a) The relatively constant level of body temperature over time
    (b) Random changes in bone thickness during the summer months
    (c) Unpredictable gains and losses of body weight throughout the year
    (d) Disordered motions of individual water molecules within the body fluids

4. The two words *dissection* and *anatomy* are practically identical in meaning, because:
    (a) Both words come from Ancient Chinese symbols
    (b) The early Romans seldom cut dead people apart
    (c) Anatomy is like physiology, while dissection is like a function
    (d) Each word contains a root for "cutting"

5. The man frequently referred to as the Father of Anatomy:
    (a) Julius Caesar
    (b) John Quincy Adams
    (c) Andreas Vesalius
    (d) George Washington

6. Anatomy is generally defined as:
    (a) Body structure and the study of body structures
    (b) Functions that occur only within the human body
    (c) The harmful alternative to physiology

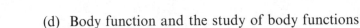

(d) Body function and the study of body functions

7. The basic characteristics of structures include:
    (a) Taking up space and relieving physical pain
    (b) Considerable mass or weight, but no particular size or shape
    (c) Possession of an underlying "skeleton" or "woven fabric" of material
    (d) Reckless abandonment of all the established Laws of Nature!

8. This sentence, "Most carbon atoms in the skin bond or connect tightly together," best represents a statement of:
    (a) Anatomy
    (b) Physiology
    (c) Plain function
    (d) Biological Disorder

9. A construction crew working on your vacant lot digs up a very large unidentified skeleton. What specific scientific specialty would probably be the most helpful in identifying it?
    (a) Embryology
    (b) Comparative anatomy
    (c) Physiology
    (d) Developmental anatomy

10. The nine levels of structural organization within the human organism can be conveniently modeled as horizontal slabs or stacks of drawers within a Great Body Pyramid. This type of geometric modeling is appropriate, since:
    (a) Many key body facts can be deposited into the square cells of a grid or matrix having a certain degree of size and complexity
    (b) Anatomy and physiology are so different from each other, that this technique is the only possible way to describe them!
    (c) Most important body facts are associated with no particular level of anatomy.
    (d) The horrible complexity of human physiology makes any attempt to classify its major facts a hopeless task!

Body-Level Grids for Chapter 1

Two key body facts were tagged with numbered icons in the page margins of this chapter. Write a short summary of each of these key facts into a numbered cell or box within the appropriate *Body-Level Grid* that appears below.

Anatomy and *Biological Order* Fact Grid for Chapter 1:

ORGAN
Level

1

Physiology and *Biological Disorder* Fact Grid for Chapter 1:

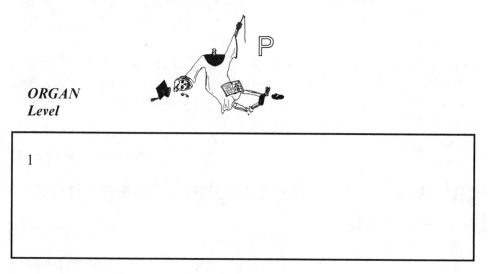

ORGAN
Level

1

CHAPTER

"Cutting Up" Human Bodyspace

In Chapter 1, we introduced the general concept of "Human Bodyspace" as a helpful model of our body's internal environment. Now, here in Chapter 2, we will begin to formally "cut up" or "dissect" Human Bodyspace, and then introduce specific terms to describe the result.

The Right Angle and Rectangle: Sacred Gifts from the Mummies!

We have seen and already discussed the Great Body Pyramids, which, with their nine levels of body organization, seem like a series of horizontal slabs or grids stacked to make a peak.

Chapter 1 pointed out how both the Ancient Egyptians and, much later, the Italian Artists "dissected" Human Bodyspace into grids or matrices having rectangular cells. But now it is time to turn away from both the pyramid and the grid, and closely examine their common structural foundation – the rectangle.

Remember that Galileo stated, "The Book of Nature is written in characters of geometry." It was probably the Ancient Egyptians, however, who played a major role in actually *creating* geometry, in the first place! The word, geometry, literally means "the process of" (*-y*) "earth" (*geo*) "measurement" (*metr*). One need only look at various *artifacts* (**AR**-tuh-facts) or "made" pieces of "art" from the early Egyptian Culture, to see the origins of geometry or "earth measurement."

When they were alive, the adults enjoyed playing *senet* (**SEN**-et). This was a game with pieces that moved over a rectangular board marked off into square cells, like a matrix (Figure 2.1, A). When they were dead, the Ancient Egyptians frequently had their *viscera* (**VIH**-ser-**ah**) or "guts" (internal organs) removed and placed into jars. Their mummified bodies were then put into a wooden *mummy case*. Each mummy case was either shaped like a rectangle, or a rectangle carved to fit the body (Figure 2.1, B). The mummy case bore numerous *hieroglyphics* (**high**-ur-uh-**GLIF**-iks). These hieroglyphics were actually "sacred carvings" that made a visual language. These sacred carvings were always framed within rectangles – arranged into either horizontal rows or vertical columns on the mummy case. Orderly rows and columns were also used as patterns for the plowing of Egyptian fields, where geometry or "earth measurement" was employed.

Finally, the mummy case was placed into a rectangular stone coffin, called a *sarcophagus* (sar-**KAHF**-uh-gus) or "flesh" (*sarc*) "eater" (*phag*).

From the above evidence, it is quite obvious that the Ancient Egyptians used rectangles both to frame their *abstract thoughts*, and to frame their *body structures (anatomy)* after they died! The word, *rectangle*, comes from the Latin for "right" (*rect*) "angle." Thus, a rectangle is a four-sided box having a right angle at each of its corners.

It was the Ancient Egyptians who first created the right angle, and they used their own *human anatomy* as their source! A man standing with one arm stretched straight out at a 90-degree angle from his body (Figure 2.2) creates a right angle. And when a vertical line is dropped from the end of his hand to the ground, the form of a rectangle is traced.

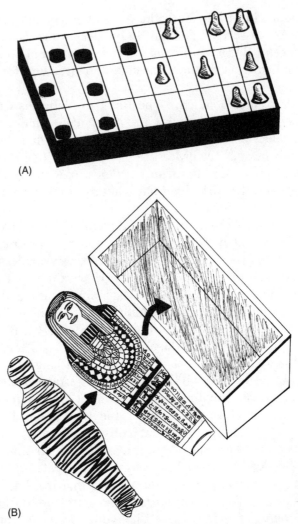

Fig. 2.1 Ancient Egypt: Rectangles in life, rectangles in death. (A) The senet board: game played on a rectangle or matrix. (B) Rectangles in death: a mummy case, hieroglyphics, and a sarcophagus or human "flesh-eater."

Anatomic Planes and Sections "Cut Up" Bodyspace

Pretend that you are holding a nice, juicy red apple in your hand. You want to cut the apple exactly in half, such that you can share it equally with a

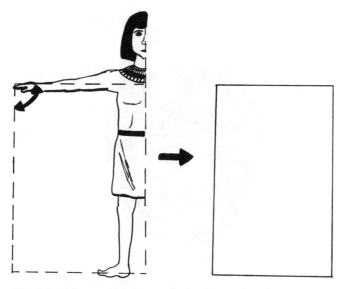

Fig. 2.2 The standing human body: Source for the rectangle.

friend. Now, assume that you want to saw a human body in half, so that it will split into right and left halves of equal size. How do you know *where* in the apple (or human body) to cut? Generally, you have to select some prominent *anatomic* (**an**-ah-**TAHM**-ik) *landmark* that marks the middle of both the apple and human "bodies." After speculating about what these landmarks might be in each case, consult Figure 2.3 for the answers.

Were you correct? For the apple, of course, you select the stem as the anatomic landmark marking the center or middle (Figure 2.3, A). Not quite so obviously, for people we select the *sagittal* (**SAJ**-ih-tul) *suture* (**SOO**-chur). In general, a suture is a "seam": that is, a zigzagging crack that marks the tight fusion of two bones. The Latin word, sagittal, literally "refers to an arrow." The sagittal suture, therefore is a jagged seam that could have been created by an arrow passing from back-to-front, scraping down over the middle of the top of the skull (Figure 2.3, B).

ANATOMIC PLANES AND SECTIONS: THE BODY'S RECTANGLES GUIDE THE CUTTING

Now that we have identified the stem of the apple, and the sagittal suture on the human skull, how do we use them to subdivide these two very different

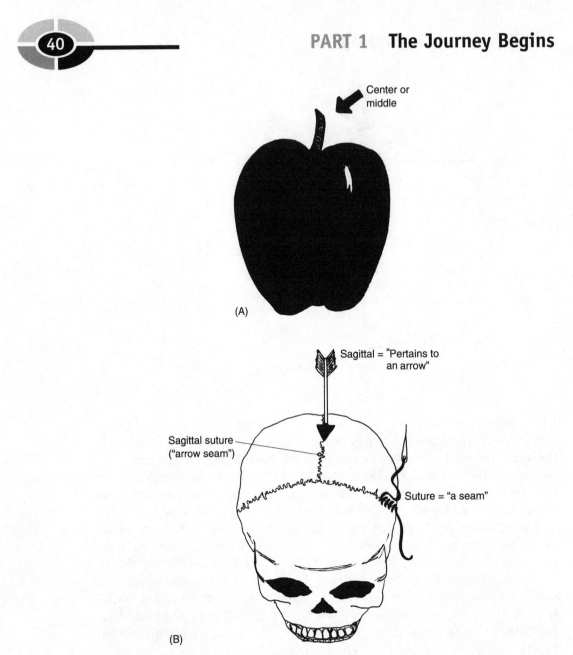

Fig. 2.3 The "bodies" of people and apples: Anatomic landmarks for the middle. (A) The stem marks the center of an apple. (B) The sagittal suture marks the middle of a skull.

types of anatomic structures? We use this information to guide the insertion of *anatomic planes* (**PLAINS**).

An anatomic plane is a flat, imaginary dividing sheet, shaped like a rectangle, that can be passed through the body at some particular angle. The anatomic plane is considered imaginary, instead of real, because it occupies only two dimensions of space (length and width), rather than all three dimensions (length, width, and thickness). Having no thickness, it is like the surface of a blackboard, sheet of paper, or a computer monitor screen.

To guide the cutting of an apple into two equal halves, an anatomic plane is passed vertically down through the stem region (Figure 2.4, A). Likewise, to guide the cutting of the human body into equal right and left halves, a plane is passed down through the middle of the sagittal suture (Figure 2.4, B). This plane is technically the *midsagittal* (**mid-SAJ**-ih-tul) *plane* or *body midline*. The midsagittal plane (body midline), therefore, is one and only one plane passed down through the middle of the sagittal suture such that it subdivides the body into exactly equal right and left halves. It is alternately called the body midline, because this plane is like a straight line, when viewed from its side.

In contrast to the single midsagittal plane, there are many possible *sagittal* or *parasagittal* (**PAIR**-uh-**saj**-ih-tul) *planes* that run parallel to and "beside" (*para-*) it. [**Study suggestion:** Why is there only one midsagittal plane, but many possible sagittal (parasagittal) planes? Explain your reasoning, then discuss it with a friend.]

As we have already noted, the anatomic plane is an imaginary rectangle passed through the body. It is valuable in helping us to do something real – actually "cut" (*sect*) the body apart into two or more pieces. This results in an *anatomic section*. An anatomic section is an actual physical cut made through the body, in the direction of some particular anatomic plane. A *midsagittal section*, for example, is an actual cut made through the body in the direction of the midsagittal plane. Like slicing an apple through its stem region, cutting a midsagittal section all the way down subdivides the body into exactly equal *anatomic right* and *anatomic left* halves.

And as Figure 2.4 clearly illustrates, such anatomic planes and sections are always passed down through the body when it is standing in the *anatomic position*. [**Study suggestion:** Using words, describe the directions of the arms, hands, feet, chin, and eyes when the human body is standing in the anatomic position.]

Terms of Relative Body Position

When the body is standing in the anatomic position, and an anatomic plane is passed through it, several different types of things can be done. One of

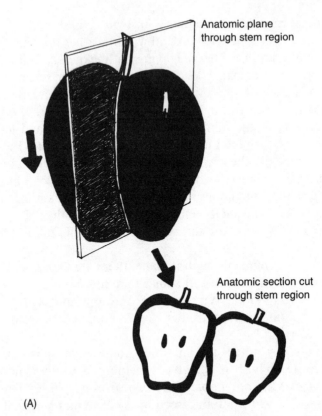

Anatomic plane
through stem region

Anatomic section cut
through stem region

(A)

Fig. 2.4 Cutting in half: The apple versus the body in its anatomic position. (A) A middle
plane and a middle section (cut) made through an apple. (B) The midsagittal plane
(body midline) and two possible sagittal (parasagittal) planes.

these is guiding the cutting of anatomic sections. Another is the creation of
terms of relative body position.

Terms of relative body position are technical words that compare the
position of one body structure to some other body structure. Anatomic
planes, in this case, are used to separate the two body structures being compared. Consider, for instance, some terms that are related to the placement of
the midsagittal plane (Figure 2.5).

One body structure is considered *medial* (**MEE**-dee-al) or *mesial* (**MEE**-
zee-al) to some other body structure, if it is closer to the "middle" (*medi-* or
mesi-) or body midline. The right eye, for example, is medial or mesial to the
right ear, because it is located closer to the body midline.

Conversely, one body structure is considered *lateral* (**LAH**-ter-al) to some
other body structure, if it is located farther to the "side" (*later*), a greater
distance away from the body midline. The right ear, for example, is lateral to

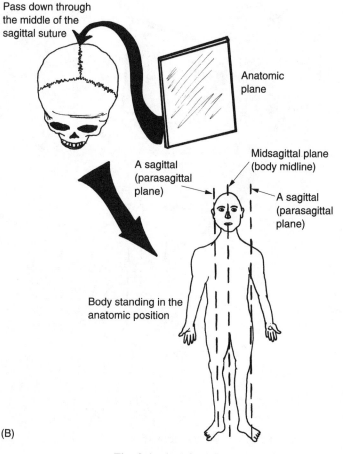

Pass down through the middle of the sagittal suture

Anatomic plane

A sagittal (parasagittal plane)

Midsagittal plane (body midline)

A sagittal (parasagittal plane)

Body standing in the anatomic position

(B)

Fig. 2.4   (continued)

the right eye, because it is located farther away from the midline, and more towards the side of the body.

TERMS RELATED TO OTHER ANATOMIC PLANES

The midsagittal plane is not the only type of anatomic plane that helps define the various terms of relative body position. Another is called a *coronal* (kor-**OHN**-al) *plane*. This plane is named after the *coronal suture* (Figure 2.6). A coronal plane is one passed down through the body in the direction of the coronal suture, which looks like a "seam" or crack made by a heavy "crown" (*coron*) smashed down hard upon a prince's head! A coronal plane is

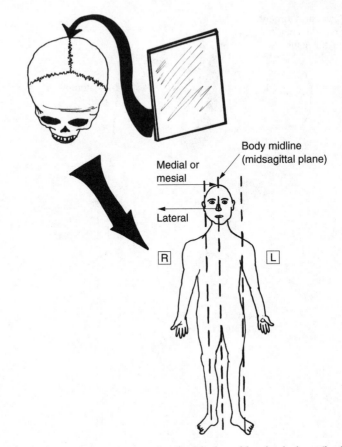

Fig. 2.5 Some important terms associated with the midsagittal plane (body midline).

alternately called a *frontal* plane, because it subdivides the body into both front and back portions.

By passing a coronal (frontal) plane between the right eye and right ear, for instance, we can say that the right eye is *anterior* (an-**TEER**-ee-er) to the right ear. This translates to mean "one that" (-*or*) is "in the front" (*anteri*). To be sure, the right eye is more "in the front" (anterior) compared to the right ear. But it is also more *ventral* (**VEN**-tral) or "pertaining to" (-*al*) the "belly" (*ventr*). For a human being standing in the anatomic position, then, the right eye is anterior or ventral to the right ear, because it is both in front of it, and closer to the belly-side of the body.

Conversely, the right ear is both *posterior* (pahs-**TEER**-ee-er) and *dorsal* (**DOR**-sal) to the right eye. In Common English, this means that the right ear is both "behind" (*posteri*) and in "back" (*dors*) of the right eye.

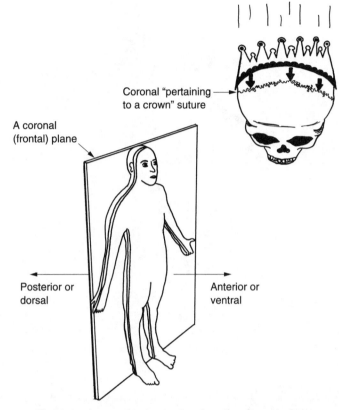

Coronal "pertaining to a crown" suture

A coronal (frontal) plane

Posterior or dorsal

Anterior or ventral

Fig. 2.6   Anatomic terms related to the coronal plane.

Transverse planes and cross-sections

A third type of anatomic plane is called a *transverse* (tranz-**VERS**) or *horizontal plane*. This anatomic plane is "turned" (*vers*) and passed "across" or "through" (*trans-*) the body in a "horizontal," left–right direction. As can be seen from Figure 2.7, a transverse (horizontal) plane subdivides the standing body into both upper and lower portions. Several terms of relative position are created via this plane. *Superior* (soo-**PEER**-ee-or), for instance, means "one that is above" (*superi*), while *inferior* (in-**FEER**-ee-or) denotes its opposite, "one that is below" (*inferi*). Synonyms (words with similar meaning) for superior are *cranial* (**KRAY**-nee-al), which "pertains to the skull" (*crani*) end of the body, and *cephalic* (seh-**FAL**-ik), which "pertains to the head" (*cephal*) end. A synonym for inferior is *caudal* (**KAW**-dal), which "pertains to the tail" (*caud*) end of the body.

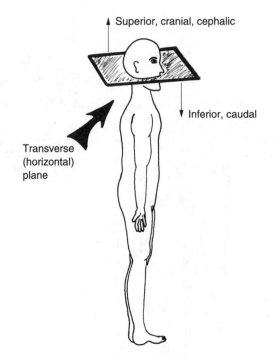

Fig. 2.7 Anatomic terms related to the transverse (horizontal) plane.

Using some specific examples, we can say that the nose is superior, cranial, or cephalic to the chin. In Common English, this means that the nose is above, more towards the skull end, or more towards the head end of the body, compared to the chin. Conversely, the chin is inferior or caudal to the nose. This means that the chin is located below the nose, or more towards the tail end of the body.

When an actual physical cut is made all the way through the body in a horizontal direction, the cut is called a *transverse section* or *cross-section*. This is because the cut is made all the way "across" the body at some particular horizontal level. Transverse (cross) sections are especially valuable in gross (macroscopic) anatomy, since they reveal many of the internal organs located within the body cavities.

TERMS NOT ASSOCIATED WITH ANATOMIC PLANES

Several terms of relative body position are not associated with any particular anatomic plane, yet still provide important information. Good examples of these are the two related terms, *proximal* (**PRAHKS**-ih-mal) and *distal* (**DIS-**

tal). Both of these terms are generally used to indicate the relative locations of various structures on the same body limb (arm or leg). "The knee is proximal to the shin." By this statement, it is meant that the knee is "closer to" (*proxim*) the origin of the leg (the hip), compared to the shin. Conversely, we have the statement, "The shin is distal to the knee." The Common English equivalent of this sentence is that the shin is "farther or distant" (*dist*) from the hip, compared to the knee.

Body Regions and Quadrants

The human body has a *central region*, which consists of the *head*, *neck*, and *trunk*:

CENTRAL REGION OF BODY = HEAD + NECK + TRUNK

The trunk, in turn, can be artificially subdivided into three main areas – the *thorax* (**THOR**-aks) or "chest," the *abdomen* (**AB**-doh-**men**) or "trunk midsection," and the *pelvis* (**PEL**-vis) or "bowl" (see Figure 2.8):

BODY TRUNK = THORAX + ABDOMEN + PELVIS

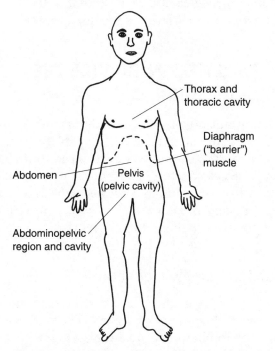

Fig. 2.8   The three main areas of the body trunk.

THE THORAX AND ITS DIAPHRAGM

The thorax (chest) is the superior portion of the trunk that lies above the *diaphragm* (**DIE**-uh-**fram**). The diaphragm is located roughly at the body "midriff," just inferior to the rib cage. It is a thin, dome-shaped sheet of muscle that forms a "barrier" between the *thoracic* (thor-**AH**-sick) *cavity* above, and the *abdominopelvic* (ab-**dahm**-ih-noh-**PEL**-vik) *cavity* below. The thoracic cavity contains the heart and both lungs.

ABDOMEN + PELVIS = THE ABDOMINOPELVIC REGION

The abdomen is the trunk midsection located caudal to the diaphragm, while the pelvis is the "bowl" (*pelv*) or *pelvic* (**PEL**-vik) *cavity* made between the two hip bones. Both on the body surface and within the body interior, there is no actual subdivision between the abdomen and the pelvis. Hence, it is proper to speak of both a combined *abdominopelvic* **region** upon the body surface, as well as an abdominopelvic **cavity** lying deep within. The abdominopelvic region is the large surface zone stretching from the diaphragm down to the *groin*s (the two places where the abdomen meets the thighs). The abdominopelvic cavity, then, is simply the large hollow space located internally within this surface abdominopelvic region.

Organ 1

ANATOMIC PLANES HELP SUBDIVIDE THE ABDOMINOPELVIC REGION

There are many important organs present within the abdominopelvic cavity. These include the liver, stomach, intestines, and spleen. From the body surface, of course, these organs cannot be directly seen. But clinical specialists, such as physicians (medical doctors) and nurses, often subdivide the abdominopelvic region into a set of four *abdominopelvic quadrants* (**KWAHD**-runts). This simple system of only "four" (*quadr*) areas helps the clinical observer localize abdominal pain and injury more effectively.

The four abdominopelvic quadrants

Let us draw a large cross right through the *umbilicus* (um-**BIL**-ih-kus), the "pit" or navel in the middle of the abdomen. The crossing of these planes results in the four abdominopelvic quadrants (Figure 2.9, A). These are called, in clockwise order, the *Right Upper Quadrant*, *Left Upper Quadrant*,

Left Lower Quadrant, and *Right Lower Quadrant*. Each of these quadrants is basically a rectangular area containing a number of viscera (guts or internal organs). Nurses and MDs often use these four quadrants to help localize the source of a patient's complaints. [**Study suggestion:** Using Figure 2.9, A, select a quadrant that would most likely be involved in a patient suffering from an attack of *acute* (ah-**KYOOT**) *appendicitis* (ah-**pen**-dih-**SIGH**-tis). Explain your thinking.]

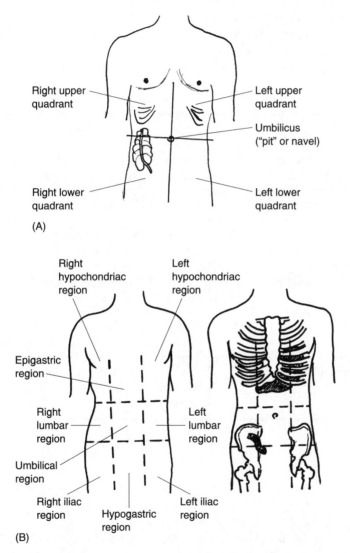

(A)

(B)

Fig. 2.9   The four quadrants and nine regions in the abdominopelvic area. (A) The four abdominopelvic quadrants. (B) The nine abdominopelvic regions.

The nine abdominopelvic regions

While clinical workers often consult the four abdominopelvic quadrants, *anatomists* (ah-**NAH**-tuh-**mists**) – "those who specialize in" (-*ist*) "anatomy" – frequently employ a more complex scheme. These are the nine *abdomino-pelvic regions* (Figure 2.9, B). Two vertical, parasagittal planes are passed just medial to the nipples. One transverse (horizontal) plane is inserted just inferior to the ribs. A second transverse plane is passed immediately superior to the hip bones. The result is a grid of nine rectangular cells, which almost looks like an Ancient Egyptian senet board (review Figure 2.1, A), drawn over the surface of the abdomen!

The nine regions are named in Table 2.1 and illustrated in Figure 2.9, B.

Table 2.1   The names and characteristics of the nine abdominopelvic regions

Region name	Important characteristics
Right and left Hypochondriac (**high**-*poh*-**KAHN**-*dree-ak*)	Both regions located "below" (*hypo-*) the "cartilage" (*chondr*) of the ribs
Epigastric (**eh**-pih-**GAS**-trik)	Lies "upon" (*epi-*) the "stomach" (*gastr*)
*Right and left Lumbar (***LUM**-*bar*)	Present in the right and left "loins" (*lumb*) or lower back area
Umbilical (**um**-**BIL**-ih-kal)	The central region including the "pit" (*umbilic*) or navel
Right and left iliac (**ILL**-ee-ak)	The two areas including much of the "flank" (*ili*) on either side of the body
Hypogastric (**high**-poh-**GAS**-trick)	The central region lying "below" (*hypo-*) both the "stomach" (*gastr*) and the umbilicus

Quiz

Refer to the text in this chapter if necessary. A good score is at least 8 correct answers out of these 10 questions. The answers are listed in the back of this book.

1. The mummy case and sarcophagus used to contain the embalmed body of an Ancient Egyptian had what main feature in common?
    (a) A heavy collection of soiled bandages
    (b) Reliance upon the rectangular form
    (c) Both always held a senet board
    (d) Scooping up piles of dirt in farmers' fields

2. The right angle has a special historical connection to human anatomy in that:
    (a) It probably originated from the tracing of a standing Egyptian with his arm extended horizontally to the ground
    (b) Geometry was used to guide the plowing and subdividing of fields
    (c) Most skull bones meet at a 90-degree angle
    (d) Humans can only stand perfectly vertical to the ground

3. Anatomic sections are different from anatomic planes, because they:
    (a) Involve no use of relative body positions
    (b) Focus chiefly upon the characteristics of physiology
    (c) Always involve the sagittal suture
    (d) Are actual physical cuts made through the body

4. The midsagittal plane is alternately known as the body midline, since:
    (a) This plane exactly marks the body middle when it is passed all the way down
    (b) It seldom intersects with the umbilicus
    (c) There are potentially an infinite number of parasagittal planes
    (d) Its position is lateral to just about everything else!

5. Pretend you are a nurse approaching the foot of a hospital bed. Your instructions are to give the patient lying in the bed a shot into the left shoulder. You then insert the needle into:
    (a) The foot directly facing you
    (b) The shoulder on your right
    (c) The shoulder on your left
    (d) Soft, fleshy tissue of both buttocks

6. "The elbow is always proximal." This statement is:
    (a) True, due to the fixed position of the elbow
    (b) False, because some people have no elbow!
    (c) True, since the elbow is always the body structure closest to the shoulder
    (d) False, since the relative position of the elbow depends upon the other body structure to which it is being compared

7. A ____ plane is the type used to subdivide the body into upper and lower portions:
   (a) Coronal
   (b) Parasagittal
   (c) Midsagittal
   (d) Transverse

8. The umbilicus is ____ to the left hip:
   (a) Posterior
   (b) Anterior only
   (c) Both medial and anterior
   (d) Distal

9. The body trunk consists of the:
   (a) Groin, neck, and gonads
   (b) Cephalic portion, alone
   (c) Most distal extremities
   (d) Thorax, abdomen, and pelvis

10. You, a physical therapist, read a clinical record of a patient. It states that, "The patient is suffering from chronic irritation in the left iliac region." It would be appropriate for you to apply hot, moist packs to what part of the patient's body?
    (a) The midriff area
    (b) Flank
    (c) Stomach
    (d) Ribs

Body-Level Grids for Chapter 2

One key body fact was tagged with a numbered icon in a page margin of this chapter. Write a short summary of this key fact into a numbered cell or box within the *Body-Level Grid* that appears opposite.

Anatomy and *Biological Order* Fact Grid for Chapter 2:

A

ORGAN
Level

1

Systemic Anatomy: A Fly-by of the Organ Systems

The first chapter in this book introduced the concept of Levels of Body Organization. You may recall that each of these levels was, in turn, visualized as a horizontal drawer or grid within a Great Body Pyramid.

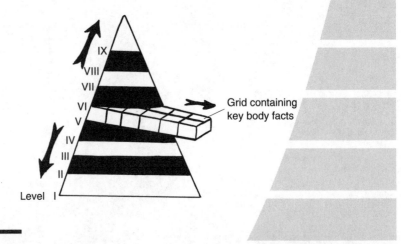

Grid containing key body facts

Level I

The Key Organ Systems Filling Human Bodyspace

The Great Body Pyramid was alternately called the internal environment or Human Bodyspace. A good way to provide an overview of Human Bodyspace is to look at some of the highest levels in the Pyramid, since these levels contain all of the lower ones. Specifically, we need to look much more closely at Level VIII – the Organ Systems.

In general, an organ system can be defined as a collection of related organs that interact together and carry out some complex body function. Before we probe ever deeper and deeper into the organs, and what lies within them, it is wise for us to provide a quick fly-by of the major organ systems in the human body.

THE INTEGUMENTARY SYSTEM – SKIN AND HAIRY CHESTS

A fine place to start our overview is with the *integumentary* (in-**teg**-you-**MEN**-tar-**ee**) *system*. The word, *integument* (in-**TEG**-you-ment), comes from the Latin for "covering." The skin, of course, being the body's covering, is the organ making up most of the integumentary system. In addition to the skin, however, are a number of *accessory* (ak-**SES**-ur-ee) *structures*, or "things added." These include the hairs, nails, and glands within the skin. The basic physiology of the integumentary system is its role in providing a boundary between the internal environment or Human Bodyspace, and the external environment surrounding it.

Organ System 1

INTEGUMENTARY SYSTEM = SKIN + ACCESSORY STRUCTURES
= SKIN + HAIRS + NAILS + GLANDS

Organ System 1

THE SKELETAL SYSTEM – HARD BONES AND NIMBLE JOINTS

If a thin person is sometimes called a "bag of bones," then if the skin or integument is the "bag," the *skeletal* (**SKEL**-uh-tal) *system* provides the "bones"! The skeletal system is a very unique organ system, in that it literally "pertains to" (-*al*) "a hard dried body" (*skelet*). Lost in the scorching vastness of the Egyptian desert, perhaps a runaway slaveworker from the Ancient Pyramids, died and collapsed in the swirling sand. With all the major viscera rotting away quickly under the baking sun, only the skin and skeleton would remain as a hard dried body!

Between the individual bone organs, one finds the *joints* or "joining" places. Unlike bones, many joints are mainly composed of *cartilage* (**CAR**-tih-luj) or "gristle." In summary of this system, we can write a simple word equation:

Organ System 2 **SKELETAL SYSTEM = BONE ORGANS + JOINTS BETWEEN THEM**

Complex functions of the skeletal system include body movement and support, as well as the important job of *hematopoiesis* (**he**-muh-toh-poi-**E**-sis) – the process of "blood" (*hemat*) "formation" (*-poiesis*).

Organ System 2 # THE MUSCULAR SYSTEM – "LITTLE MICE" HANGING ON OUR BONES

If you would be brave (or foolish) enough to cut through the skin very deeply, you would see over 600 "little mice" hanging onto the bones of your skeleton! These are the individual *skeletal muscle organs*. They make up the bulk of the *muscular* (**MUS**-kyoo-lar) *system*. Reflecting our amusing imagination, the word, muscular, "pertains to a little mouse" (*muscul*).

A glance down at Figure 3.1 reveals the source behind this highly visual meaning of the Early Greek scholars. Consider a very familiar skeletal muscle

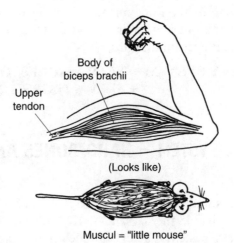

Body of
biceps brachii

Upper
tendon

(Looks like)

Muscul = "little mouse"

Fig. 3.1   The biceps brachii muscle and one of its tendons as a "little mouse" below the skin.

organ, the *biceps* (**BUY**-seps) *brachii* (**BRAY**-kee-**eye**) muscle in the upper "arm" (*brachii*). The *body* or main mass of the biceps brachii does, indeed, closely resemble a "little mouse," doesn't it?

And how about those *tendons* (**TEN**-duns) or tough straps of *connective tissue*? All of the skeletal muscles are anchored to the bones at either end by one or more of these tough, slender tendons. Doesn't one of the tendons of the biceps brachii shown in Figure 3.1 look a lot like a tail of a little mouse (muscul) creeping below the skin?

MUSCULAR SYSTEM = SKELETAL MUSCLE ORGANS + TENDONS

Organ System 3

Every time the skeletal muscles shorten or *contract*, they pull upon their tendons. The tendons in turn pull upon the bones. And the bones, in turn, move the body at their joints. The muscular system, therefore, makes up the *active* system for body movement, which creates the pulling force that eventually moves the bones.

THE NERVOUS SYSTEM – JOLTS US INTO ACTION!

Organ System 3

"Okay," you may be responding at this point. "I see that it is the skeletal muscles that contract, thereby pulling upon the bones. But *what* is it that excites the skeletal muscles to shorten or contract in the first place?" The answer to this question is a simple one – the *neurons* (**NUR**-ahns) or nerve cells present within the *nervous* (**NER**-vus) *system*. Figure 3.2 shows how a particular type of nerve cell, called the *motor neuron*, releases molecules that excite a *skeletal muscle fiber* to contract.

Some of these motor neurons are present within the *brain*, while others are located in the *spinal cord*. And after the spinal cord, various *nerves* supply the skeletal muscles.

Besides stimulating muscles to contract, there are other nerves that supply *sensory* (**SEN**-sor-ee) information about the body – such as pain, touch, and temperature – back towards the brain and spinal cord. Such information comes from *sensory receptors* (ree-**SEP**-tors) or "receivers." This is a vital communication function that helps regulate the internal environment.

Organ System 4

Overall, then, we have the following summary equation:

NERVOUS = BRAIN + SPINAL CORD + NERVES + SENSORY
SYSTEM   RECEPTORS

The nervous system is one of the body's major systems for communication and control of the internal environment.

Organ System 4

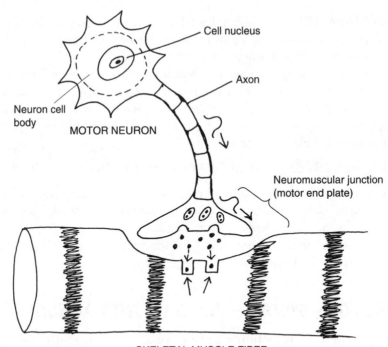

Fig. 3.2   A motor neuron excites a skeletal muscle fiber.

THE GLANDULAR SYSTEM – OUR SERVANTS OF SECRETION

Organ System 5

In addition to the nervous system, the body contains an extensive *glandular* (**GLAN**-dyoo-lar) *system* – a bunch of "little acorns" or *glands*. These glands are actually fairly rounded, "acorn"-shaped masses of cells specialized for the function of *secretion* (sih-**KREE**-shun). Secretion is literally the "process of separating" – the separation of certain substances from the bloodstream, followed by their release. Secretion is essentially the release of some useful substance that performs a needed function within the body. One of the broad functions of secretion is helping the nervous system in its job of communication and control of the internal environment.

There are dozens of glands, and each of them may be considered an organ specialized for secretion. For example, use your fingers to *palpate* (**PAL**-payt) or "gently touch" the *thyroid* (**THIGH**-royd) *gland* located on the anterior (front) surface of your neck. The thyroid gland is one type of *endocrine* (**EN**-doh-krin) *gland* – a gland of "internal" (*endo-*) secretion of a *hormone* (chemical messenger) directly into the bloodstream. [**Study suggestion:** Can you name one specific hormone secreted by the thyroid gland? Try to answer this question, then check with Chapter 11.]

The other major type of glands are the *exocrine* (**EK**-suh-krin) *glands* or glands of "external" (*exo-*) secretion of some useful product into a *duct*. Consider, for instance, the hundreds of tiny *sweat glands* embedded within your skin. Each of them secretes *sweat* into its overlying *sweat duct*, which then carries the sweat outwards, towards the skin surface. The sweat secretion is useful, of course, because it helps cool the body and maintain homeostasis or relative constancy of body temperature.

Endocrine versus exocrine glands are briefly contrasted within Figure 3.3.

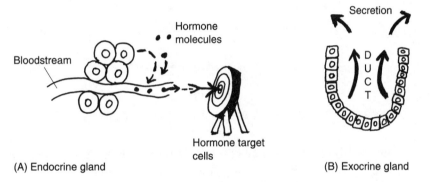

(A) Endocrine gland   (B) Exocrine gland

Fig. 3.3 Two major types of gland organs: Endocrine versus exocrine.

GLANDULAR = **ENDOCRINE GLANDS** + **EXOCRINE GLANDS**
SYSTEM   (*Internal secretion*   (*External secretion*
         into bloodstream)   *into ducts*)

Organ System 5

THE CIRCULATORY SYSTEM – BLOOD THROUGH PUMPS AND PIPES

The *circulatory* (**SIR**-kyuh-luh-**tor**-ee) *system* is alternately known as the *cardiovascular* (**car**-dee-oh-**VAS**-kyuh-lar) *system*. The circulatory (cardiovascular system) is the organ system containing the "heart" (*cardi*) and "little vessels" (*vascul*). It is called circulatory because the blood tissue does, in fact, travel in a "little circle" (*circul*), both beginning and ending with the heart. (Examine Figure 3.4.)

CIRCULATORY = **THE HEART** + **BLOOD** + **BLOOD**
(CARDIOVASCULAR)   **ORGAN**   **VESSEL**   **TISSUE**
SYSTEM   **ORGANS**

Organ System 6

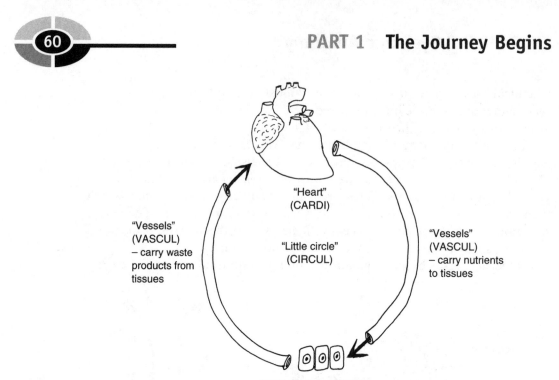

"Heart"
(CARDI)

"Little circle"
(CIRCUL)

"Vessels"
(VASCUL)
– carry waste
products from
tissues

"Vessels"
(VASCUL)
– carry nutrients
to tissues

MAIN BODY TISSUES

Fig. 3.4   An overview of the circulatory (cardiovascular) system.

Organ System 6

The complex function carried out is the temporary storage of blood and its contained *nutrients* (**NEW**-tree-ants) and waste products within the heart, followed by its circulation throughout the body by means of the blood vessels.

THE LYMPHATIC-IMMUNE SYSTEM – PROTECTION WITH "CLEAR SPRING WATER"

Closely associated with the circulatory (cardiovascular) system is another critically important organ system, the combined *lymphatic* (lim-**FAT**-ik) and *immune* (ih-**MYEWN**) *system*. The word, *lymph* (**LIMPF**), means "clear spring water," while lymphatic "pertains to lymph or clear spring water." Lymph essentially consists of the clear, watery fluid that is filtered out of the *blood capillaries* (**CAP**-uh-**lair**-ees), the tiniest blood vessels. As shown in Figure 3.5, the clear material filters out of the blood capillaries, and into nearby *lymphatic capillaries*. These lymphatic capillaries are part of a much larger *lymphatic circulation*, which consists of vessels that mainly run parallel to the blood vessels. Some of the vessels of the lymphatic circulation pass through the *lymphatic organs*, which help cleanse the lymph.

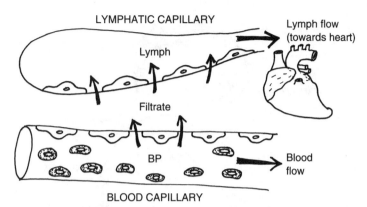

Fig. 3.5   Blood capillaries filter lymph into the lymphatic system. BP = blood pressure.

The word, immune, literally means "not serving (disease)." The immune system, therefore, is an organ system that does "not serve (disease)," because it helps provide protection from it. Specifically, the immune system produces *antibodies* (**AN**-tih-**bah**-dees), chemicals that destroy foreign invaders, such as deadly *bacteria* (back-**TEER**-ee-uh). Because the lymph usually contains these chemical antibodies (and other protectors from disease), the lymphatic and immune systems are often combined together as a single lymphatic-immune system.

Organ System 7

LYMPHATIC-IMMUNE =	LYMPHATIC	+	IMMUNE
SYSTEM	SYSTEM		SYSTEM
	(Lymphatic vessels		*(Antibodies and other*
	and organs)		*protectors from disease)*

Organ System 7

THE RESPIRATORY SYSTEM – AIR THROUGH BAGS AND PIPES

The circulatory and lymphatic-immune systems are not the only organ systems in the human body that involve a series of hollow tubes. Important to include in this group is the *respiratory* (**RES**-pir-ah-**tor**-ee) *system*. As its name states, the respiratory system is the organ system responsible for *respiration* (res-pir-**AY**-shun) – "the process of" (*-tion*) "breathing" (*spir*) "again" (*re-*).

When a person breathes again and again, the air is *inhaled* (**IN**-hailed) *into* the lungs, then *exhaled* (**EKS**-hailed) *out* of the lungs (see Figure 3.6).

Anatomically speaking, the respiratory system consists of the *Upper Respiratory Pathway* plus the *Lower Respiratory Pathway*. The Upper Respiratory Pathway extends from the nose and mouth, all the way down

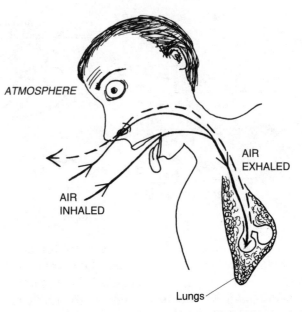

ATMOSPHERE

AIR
EXHALED

AIR
INHALED

Lungs

Fig. 3.6 The respiratory system: Organs for "breathing again."

Organ System 8

to the *trachea* (**TRAY**-kee-ah) or main "windpipe." The Lower Respiratory Pathway, in contrast, starts from the *right* and *left primary bronchi* (**BRAHN**-kigh), the major branches of the trachea, and extends deep down into the lungs.

Physiologically speaking, the respiratory system is critical for the key function of *oxygenation* (**ahks**-ih-jen-**AY**-shun) – the "process of" (-*tion*) providing fresh "oxygen" to the body tissues. It also helps rid the body of *carbon dioxide* (die-**AHKS**-eyed), a major waste product, as well as aid in maintaining body *acid–base balance*.

Organ System 8

RESPIRATORY =	UPPER RESP.	+	LOWER RESP.
SYSTEM	PATHWAY		PATHWAY
	(*Nose/mouth to trachea*)		(*Bronchi through both lungs*)

THE DIGESTIVE SYSTEM – OUR BODY'S "GRINDER"!

As soon as you start talking about *digestion* (die-**JES**-chun), you literally start talking about "dividing or dissolving." Hence, we might picture the *digestive* (die-**JES**-tiv) *system* as a long, winding meat-grinder (you know, a metal one you grind by hand with a crank!). Food is taken in through the *oral* (**OR**-al) *cavity* or "mouth," and then it is divided or ground-up into ever-smaller pieces.

The resulting *nutrients* (**NEW**-tree-unts) are important "nourishing (substances)" that provide the body with its needed energy. Such vital nutrients include *glucose* (**GLUE**-kohs) and other simple sugars, *lipids* (**LIP**-ids) or fatty substances, as well as *proteins* (**PRO**-teens).

Many of the *digestive organs* help in this process of dividing or dissolving consumed food and liquid into its component nutrients. Likewise, many of them are involved in the process of producing waste products, which are eventually eliminated from the body as *feces* (**FEE**-sees).

The digestive system begins with the oral cavity and ends with the *anus* (**AY**-nus). Overall, its chief functions are digestion of foodstuffs, absorption of nutrients, and elimination of feces.

Organ System 9

DIGESTIVE SYSTEM = ORAL CAVITY DOWN TO ANUS

Organ System 9

THE GENITOURINARY SYSTEM – KEEPER OF OUR URINE AND SEX!

The final organ system for our review is the *genitourinary* (**JEN**-ih-toh-**ur**-ih-**nair**-ee) or *urogenital* (**UR**-oh-**jen**-ih-tal) *system*. These alternate names reflect the fact that this system really represents the combination of two others.

The *urinary* (**UR**-ih-**nair**-ee) *system*, of course, is the organ system responsible for the production and excretion of urine. It also assists in maintaining both *body salt-and-water balance*, as well as acid–base balance.

The *genital* (**JEN**-ih-tal) *system* is alternately known as the *reproductive* (**ree**-proh-**DUCK**-tiv) *system*. The term, genital, "refers to begetting or producing." In general, this is the job of the *genital organs* in both males and females. The genital organs, such as the *penis* (**PEA**-nis) in males and the *vagina* (vah-**JIE**-nuh) in females, are used to "beget or produce" new children, sexually. This is where the term, reproductive, comes into play. The reproductive system in the male and female serves to "produce" new human organisms "again" (*re-*).

"Okay," you may now be wondering, "but why are the reproductive and urinary systems classified together as a single genitourinary (urogenital) system?" The reason is that many portions of these two systems occur together within the *same* organs! The penis, for instance, functions to both excrete urine (making it a urinary organ) and release *sperm cells* (making it a genital or reproductive organ). Hence, the penis is properly characterized as neither a purely urinary organ, nor a purely genital or reproductive one. Rather, it is a genitourinary (urogenital) organ.

Organ System 10

$$\begin{array}{ccc} \text{GENITOURINARY} & = & \text{GENITAL OR} & + \text{URINARY} \\ \text{(UROGENITAL) SYSTEM} & & \text{REPRODUCTIVE ORGANS} & \text{ORGANS} \end{array}$$

What Are "Regional" and "Surgical" Anatomy?

A

Organ System 10

Up to this point, we have been classifying various organs of the human body into particular inter-related groups of organs called the organ systems. There is also, however, a so-called *Regional or Surgical* (**SUR**-jih-**kal**) *Anatomy*.

When surgeons operate on a damaged part of the human body, the operation usually only covers a particular *region* of the body, such as the shoulder, hip, leg, abdomen, or chest. Therefore, in regional (surgical) anatomy, the surgeon wants to know what particular blood vessels, nerves, muscles, and other organs are located *just within that particular body area*. This approach, then, cuts across the traditional boundaries between the classified organ systems. Instead, surgical or regional anatomy emphasizes the *local* (regional) body structures of interest to the operation, whatever these might be.

Pathological Anatomy Creates Pathophysiology: It's All About "Suffering"

In discussing regional or surgical anatomy, it is important to realize just *why* this very functional and practical, surgical-operation based, type of anatomy exists in the first place. The reason for its existence is the common occurrence of *pathological* (**path**-uh-**LAHJ**-uh-kul) *anatomy*.

The concept of Biological Disorder has already been introduced (Chapter 1). Recall that Biological Disorder essentially involves a break in normal body patterns. A case was made, for instance, about a fractured long bone. Further, Chapter 1 pictured broken patterns of structure as existing at different levels of body organization, within the Great Pyramid of Structural Disorder.

This thinking can now be extended with the idea of pathological anatomy. The word, *pathology* (path-**AHL**-uh-jee), exactly translates to mean "the study of" (*-ology*) "suffering" (*path*). Suffering, of course, is closely related to disease. Pathological anatomy, therefore, can be defined as the study of body structures having broken patterns, which leads to a condition of suffering or disease. Normal anatomy, in marked contrast, involves the study of body structures having intact patterns, which leads to a condition of *Clinical* (**KLIN**-ih-kal) *Health*. By clinical health, we mean that there is no reason to visit a "clinic" or physician's office, so that the person must be healthy!

NORMAL = BIOLOGICAL ORDER = A STATE OF
ANATOMY   (INTACT PATTERNS OF   CLINICAL HEALTH
BODY STRUCTURE)

versus

PATHOLOGICAL = BIOLOGICAL DISORDER = A STATE OF
ANATOMY   (BROKEN PATTERNS OF   "SUFFERING"
BODY STRUCTURE)   OR DISEASE

Since states of Biological Disorder are also generally reflected in states of "suffering" or disease, the same icon (a fallen-down and fractured Professor Joe skeleton) will be used for examples of pathological anatomy. In practical use, however, pathological anatomy is often discovered in samples of either gross (macroscopic) anatomy, or in samples of microscopic anatomy, that are obtained during either *tissue biopsy* (**BUY**-ahp-see) or whole-body *autopsy* (**AW**-tahp-see).

Tissue biopsy is literally "a vision of" (*-opsy*) "life" (*bio*). A sample of cells or tissues is removed from a living patient, and is then examined under the microscope for various key abnormalities in structure that are associated with Biological Disorder. If such structural abnormalities are discovered, then the tissue sample is said to show pathological anatomy. The physician then usually *diagnoses* (**die**-uhg-**KNOW**-sus) some disease in the patient.

Autopsy is literally "a vision of" (*-opsy*) a dead body or cadaver by one's own "self" (*auto-*). As pioneered by Andreas Vesalius (Chapter 1), autopsies of cadavers are intended to reveal possible Biological Disorders (pathological anatomy) of grossly visible internal organs. The presence of a large hole in the wall of the heart's left ventricle, for instance, could be seen during autopsy and listed as a major possible cause of a patient's death. Such a large, abnormal hole would certainly be classified as pathological anatomy!

Pathophysiology

Closely following from pathological anatomy is *pathophysiology* (**path**-oh-**fizz**-ee-**AHL**-uh-jee). Recall that *path* means "diseased." It also indicates "suffering." If particular body structures have pathological anatomy, then their associated functions are usually highly abnormal as well. As a result, the person is "diseased and suffering." Let us return to the mention of a large, abnormal hole seen in the left ventricle during autopsy. This is a dramatic example of pathological anatomy that would create severe pathophysiology and abnormalities in the pumping actions of the heart. The result is *morbidity* (mor-**BID**-ih-**tee**) – "a condition of" (-*ity*) "illness" (*morbid*).

PATHOLOGICAL ANATOMY	**Generally results in**	**PATHOPHYSIOLOGY (Abnormal body functions associated with disease)**

The other companion volume to this "Dynamic Duo," *PHYSIOLOGY DEMYSTIFIED*, discusses fascinating topics in pathophysiology in much greater detail.

Quiz

Refer to the text in this chapter if necessary. A good score is at least 8 correct answers out of these 10 questions. The answers are listed in the back of this book.

1. An organ system is:
   (a) Seldom involved in important body functions
   (b) A group of related organs that carry out some complex body function
   (c) About 1/3 to 1/2 of all normal structures in the Great Pyramid
   (d) All of the more than two dozen organs within the entire body

2. The skeletal system:
   (a) Includes various accessory structures, such as nails
   (b) Consists of the bones in the body limbs and trunk
   (c) Frequently is interrupted by damage to blood tissue
   (d) Consists of the joints between the bone organs, as well as the bones

3. The —— consists of over 600 "little mice":
   (a) Integumentary system
   (b) The collection of tendons and joints
   (c) Sweat glands that secrete "cheesy"-smelling sweat!
   (d) Muscular system

4. The sensory receptors are classified as members of which organ system?
   (a) Nervous
   (b) Endocrine
   (c) Epithelial
   (d) Genitourinary, only

5. The circulatory system:
   (a) Collects and distributes the lymph
   (b) Stores and pumps the blood in a "little circle"
   (c) Absorbs body nutrients
   (d) Functions completely independently of the heart

6. The antibodies are found within the:
   (a) Lymphatic-immune system
   (b) Respiratory system
   (c) Digestive system
   (d) Nervous system

7. The ____ is involved in "begetting or producing":
   (a) Glandular system
   (b) Integumentary system
   (c) Genitourinary system
   (d) Muscular system

8. Tendons are most closely connected to the:
   (a) Integumentary system
   (b) Muscular system
   (c) Glandular system
   (d) Skeletal system

9. A so-called "successful" autopsy would most likely find evidence of:
   (a) Tissue growth
   (b) Pathological anatomy
   (c) Normal anatomy
   (d) Macroscopic anatomy

10. A surgeon spends considerable time in memorizing and visualizing the blood vessels, bones, nerves, muscles, and joints in the knee. This shows the practical importance of knowledge in:
   (a) Microscopic anatomy
   (b) Comparative body structures
   (c) Human physiology
   (d) Regional anatomy

Body-Level Grids for Chapter 3

Several key body facts were tagged with numbered icons in the page margins of this chapter. Write a short summary of each key fact into a numbered cell or box within the *Body-Level Grids* that appear below.

Anatomy and **Biological Order** Fact Grids for Chapter 3:

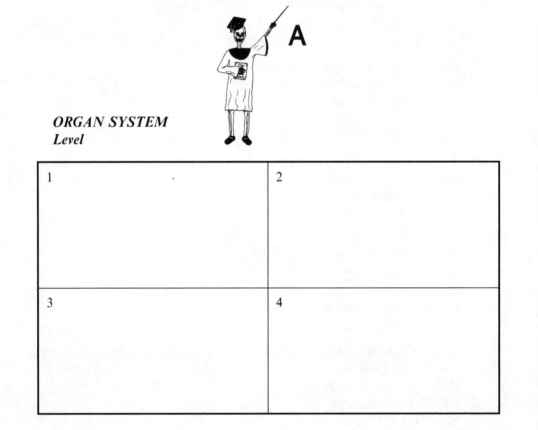

ORGAN SYSTEM
Level

1	2
3	4

ORGAN SYSTEM
Level

5	6
7	8

9	10

Physiology and *Biological Order* Fact Grids for Chapter 3:

ORGAN SYSTEM
Level

1	2
3	4

ORGAN SYSTEM
Level

5	6
7	8

ORGAN SYSTEM
Level

9	10

Test: Part 1

DO NOT REFER TO THE TEXT WHEN TAKING THIS TEST. A good score is at least 18 (out of 25 questions) correct. Answers are in the back of the book. It's best to have a friend check your score the first time, so you won't memorize the answers if you want to take the test again.

1. The forearm bones are followed by the wrist bones, and the wrist bones are followed by the finger bones. This situation provides a good example of:
   (a) Broken body patterns
   (b) Homeostasis
   (c) Biological Order
   (d) Normal physiology
   (e) Plain body functions

2. A shoe flies off one hoof of a galloping horse. It strikes the side of a red barn. This accident is an illustration of:
   (a) Anatomy
   (b) Physiology
   (c) Biological Order

(d) Dissection

(e) Complementarity

3. The word root (main idea), *tom*, exactly translates to mean:
   (a) "Body structure"
   (b) "Comparative anatomy"
   (c) "Body function"
   (d) "Cut"
   (e) "Removal of"

4. Bodies that have "fallen dead":
   (a) Cadavers
   (b) Torsos
   (c) Organ systems
   (d) Plain functions
   (e) Physi's

5. "The skull of a human typically has a greater width and capacity than does the skull of an ape with a body of equal size." This statement is most closely associated with:
   (a) The World of Physiology
   (b) Comparative anatomy
   (c) The cellular level
   (d) Photosynthesis
   (e) Glucose molecules

6. The highest of the nine levels of body organization is:
   (a) Atoms
   (b) Tissues
   (c) Organ systems
   (d) Organism
   (e) Cell

7. Who wrote that, "The Book of Nature is written in characters of Geometry":
   (a) Galileo
   (b) Andreas Vesalius
   (c) Albrecht Durer
   (d) Baby Heinie
   (e) Leonardo da Vinci

8. An anatomic plane does not fit the definition of a structure, since it:
   (a) Has great mass
   (b) Has only length and width

   (c) Carries out a great many body functions
   (d) Can be subdivided into a mesh or grid lattice
   (e) Frequently has pretty colors

 9. Gross or macroscopic anatomy:
   (a) Involves only body structures that are clearly visible to the unaided eyes
   (b) Never involves human embryos, which are always tiny
   (c) Centers most of its attention upon human and animal cells
   (d) Really is just another name for comparative anatomy
   (e) Is identical to pathological anatomy

10. The Ancient Egyptians probably played a crucial role in anatomy by:
   (a) Eliminating religious ideas from the understanding of body functions
   (b) Making mummies look good enough for display in future museums
   (c) Creating the concepts of right angles based upon the standing human form
   (d) Subdividing the desert into gridded plots for farming
   (e) Irrigating the Nile River for human water consumption

11. The sagittal suture:
   (a) Passes across the skull like a crown
   (b) Forms a right angle with all known bony landmarks
   (c) Provides a landmark for identifying the body midline
   (d) Represents a highly movable joint
   (e) Frequently shifts its position within adults

12. An anatomic section is:
   (a) An imaginary product of the human mind
   (b) The same thing as an anatomic plane
   (c) Seldom seen in anatomy classes
   (d) Is usually made only distal to the elbow
   (e) A real physical cut actually sliced through part of the body

13. The right nipple would be considered ____ to the breastplate:
   (a) Medial
   (b) Superior
   (c) Lateral
   (d) Posterior
   (e) Internal

14. A coronal plane has an alternate name, the:
    (a) Frontal suture
    (b) Dorsal nerve tract
    (c) Coronal suture
    (d) Frontal plane
    (e) Posterior aspect

15. The cephalic direction is the exact opposite of what direction?
    (a) Caudal
    (b) Cranial
    (c) Medial
    (d) Dorsal
    (e) Lateral

16. The elbow is ____ to the wrist:
    (a) Proximal
    (b) Transverse
    (c) Inferior
    (d) Posterior
    (e) Distal

17. BODY ____ = THORAX + ABDOMEN + PELVIS
    (a) HEAD
    (b) QUADRANT
    (c) TRUNK
    (d) TOES
    (e) MIDRIFF

18. The four abdominopelvic quadrants can be identified by drawing a cross through the:
    (a) Left salivary gland
    (b) Diaphragm muscle
    (c) Bony pelvis
    (d) Umbilicus
    (e) Lower groin

19. Abdominopelvic region located "below" the rib "cartilage":
    (a) Hypochondriac
    (b) Umbilical
    (c) Iliac
    (d) Epigastric
    (e) Hypogastric

20. ___ SYSTEM = SKIN + ACCESSORY STRUCTURES
    (a) RESPIRATORY
    (b) PROPER PH
    (c) SKELETAL
    (d) INTEGUMENTARY
    (e) MUSCULAR

21. Motor neurons in the brain are part of what organ system?
    (a) Gonadal
    (b) Nervous
    (c) Liver
    (d) Digestive
    (e) Muscular

22. The two main types of glands in the glandular system:
    (a) Sympathetic and parasympathetic
    (b) Proximal and distal
    (c) Salivary and reproductive
    (d) Endocrine and vascular
    (e) Exocrine versus internally secreting glands

23. Blood flowing through its vessels anywhere in the body would be included as part of the:
    (a) Lymphatic system
    (b) Digestive tract
    (c) Cardiovascular system
    (d) Respiratory tree
    (e) Urinary calculi

24. What emphasizes local rather than systemic groupings of body structures?
    (a) Circulatory dynamics
    (b) Surgical anatomy
    (c) Vertebrate physiology
    (d) Pathological anatomy
    (e) Embryology

25. A state of Clinical Health is most closely linked with:
    (a) The Humoral Doctrine
    (b) Biological Order in both anatomy and physiology
    (c) Pathological anatomy
    (d) Pathophysiology
    (e) Biological Disorder in either anatomy or physiology

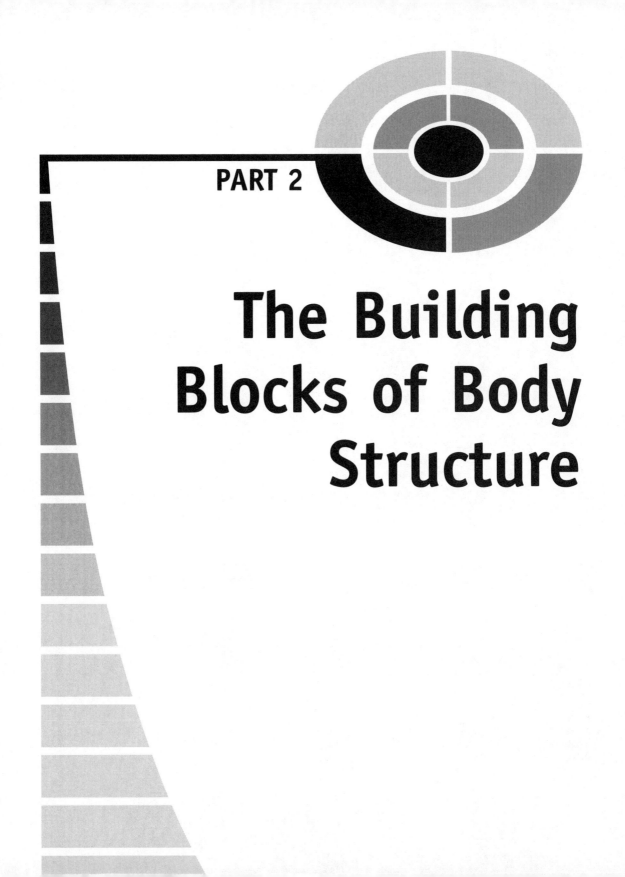

PART 2

The Building Blocks of Body Structure

Hidden Chemicals in Our Body Basement: The Great Pyramid Starts Out as a Midget!

In Part 1 of this book, we presented the major themes for further development in *ANATOMY DEMYSTIFIED*. One of the most important of these themes was the idea of Levels of Body Organization. Chapter 3, Systemic Anatomy, briefly outlined the organ systems. You may remember these as located near the top or apex of the Great Body Pyramid.

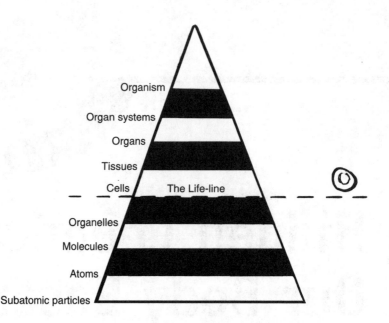

The Chemical Level: Looking for Dirt in the Body Basement

The Organ System (Level Number VIII) is one of the most prominent and directly observable features of the Pyramid. Like the front of our own house, we want our Integumentary System, for example, to have real "curb appeal" for all casual onlookers. This same understandably human tendency to emphasize external appearances leads us to stand over a fresh cadaver in a casket and exclaim, "But he (or she) looks so *good*!"

Due to the skilled *cosmetic* (kahs-**MET**-ik) or "orderly arranging" efforts of a *mortician* (mor-**TISH**-un) – "one who specializes in" (-*ician*) fixing up the "dead" (*mort*) – the expired person has a pleasing appearance to help comfort us in our time of sorrow. Yet, we very much need to realize the truth of the old saying that, "Beauty is not just skin deep."

ANATOMY BEGINS IN THE HIDDEN WORLD OF ATOMS

When the eye lightly scans over the gross *superficial* (soo-per-**FISH**-al) aspect of the body "surface," it completely misses the billions and billions of elegant structures that are far too microscopic to be seen. These Unseen Billions of tiny particles make up the Hidden World of Atoms. An *atom* (**AH**-tum) is the simplest form of a chemical *element* – a primary type of matter. The four most common elements (atoms) found in the human body are carbon (C), oxygen (O), hydrogen (H), and nitrogen (N).

These four types of atoms have the shape of round particles or spheres, resembling tiny particles of dirt. Like pieces of dirt swept under the skin carpet (where no one can see them), the Hidden World of Atoms provides a broad, deep foundation upon which the entire Body House is built. To fully understand human anatomy, therefore, we must begin at the *Chemical Level* of body organization.

The Chemical Level consists of the three bottom levels of the Great Body Pyramid (Figure 4.1). These are the subatomic particles, atoms, and molecules.

Atom 1

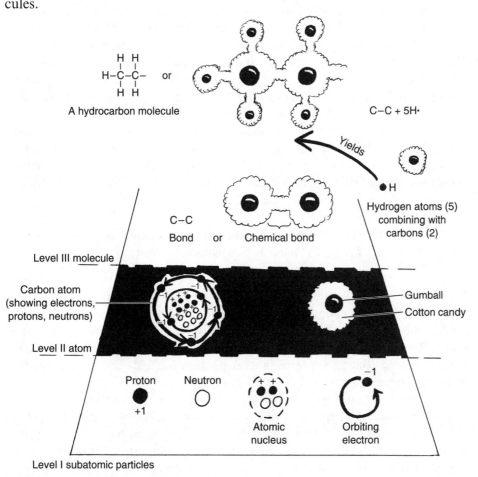

Fig. 4.1   The chemical level of body organization.

THE CHEMICAL = SUBATOMIC + ATOMS + MOLECULES
       LEVEL   PARTICLES

A

Subatomic
particles 1

Atoms and their subatomic particles

Occupying Level I, "below" (sub-) the "atom," at the very base of the Pyramid, are the subatomic particles. The three main types of subatomic particles are the *protons* (**PROH**-tahns), *neutrons* (**NEW**-trahns), and *electrons* (e-**LEK**-trahns). These subatomic particles are "below" (sub-) the level of the *whole* atom, in the sense that they are the major pieces *making up* the atom.

Each atom, such as the C (carbon) atom, contains a central "kernel" called the *nucleus* (**NEW**-klee-**us**). [**Study suggestion:** Visualize the nucleus as a hard, round gumball, which can be cut into other rounded particles of still smaller size.] There are two types of subatomic particles found within the *atomic* (ah-**TAHM**-ik) nucleus – the protons and the neutrons. Each proton has a net (overall) electrical charge of +1. Each neutron, however, has zero net charge, such that it is electrically "neutral." The smallest atom, hydrogen (H), contains only a single proton (and no neutrons) within its nucleus.

Rapidly orbiting around the nucleus, at various distances, are the electrons. Each of these particles has a net charge of −1. The electrons may be thought of as a cloud of negative charge, surrounding the central nucleus. [**Study suggestion:** As shown in Figure 4.1, it is helpful to imagine the electron cloud as a sticky puff of cotton candy, around a hard gumball nucleus.]

SUBATOMIC = (Within the Nucleus): + (Around the Nucleus):
PARTICLES   Protons + Neutrons   Clouds of Electrons

A

Subatomic
particles 2

You might now be wondering to yourself: "Well, if subatomic particles are so small, can they *still* even be considered body structures or anatomy, *at all?*" Technically speaking, yes! This is especially true for the protons and neutrons. Remember (Chapter 1) that one of the key characteristics of all structures is having some mass or weight. A proton, for instance, has a mass that is 1,836 times as much as that of an electron.

A

Atom 2

The carbon atom generally has a nucleus with 6 protons and 6 neutrons. Because it also has 6 negatively-charged electrons orbiting around its nucleus, the +6 charge of the nucleus is exactly counterbalanced by the −6 charge of all the electrons. The carbon atom, like all regular atoms, thus has a net electrical charge of $6 - 6 = 0$. Here we have, then, a fine example of balance and order within the Hidden World of Atoms.

Molecules = Strings of Bonded Atoms

It is certainly accurate to say that the overall anatomy of the human organism ultimately comes down to the structures of all the billions of unseen atoms and their subatomic particles within the body. The carbon atom, which we have pictured back in Figure 4.1, is basically a tiny black sphere. But, from the Molecule Level (III) on up, the characteristics of the body structures encountered depend more on the *interactions* and *combinations* between various atoms, rather than the atoms themselves. The human stomach, for instance, looks rather glistening and pinkish in color, when viewed during an operation of the abdomen. Yet, much of the structure of the stomach wall ultimately does come down to its billions of carbon atoms. "Okay, you may now wonder, then why doesn't the wall of the stomach *look* black, since it *contains* so many *black-colored* carbon atoms?"

MOLECULES

The answer to the above question is simply, "Because the stomach wall contains carbon and other atoms linked into many *different* kinds of *molecules,* most of which are *not* colored black!"

A molecule is a combination of two or more atoms held together by *chemical bonds*. A chemical bond is a linkage between the outer electron clouds of different atoms. In some chemical bonds, the electron clouds are shared fairly equally between the atoms. In others, the clouds are shared unequally, or are even completely transferred from one atom to another.

A

Molecule 1

Looking back at Figure 4.1 one more time, we can see a C–C (carbon–carbon) chemical bond, in which the electron clouds are shared equally. [**Study suggestion:** Imagine two sticky puffs of cotton candy jammed together, then pulled slowly apart, with equal force at either end. The resulting carbon–carbon bond has an equal amount of cotton candy electron cloud around each nuclear gumball.]

This can be contrasted with an H–C or hydrogen–carbon chemical bond. When several hydrogen–carbon bonds are made, a *hydrocarbon* (**HIGH**-droh-**kar**-bun) molecule is created. Such hydrocarbon molecules can involve dozens of carbon atoms, and hundreds of hydrogen atoms! A huge number of H–C (hydrocarbon) bonds are often present within the body *macromolecules* (**MAK**-roh-**mall**-uh-kyewls): that is, the very "big" (*macr*) ones! Such body macromolecules include the *DNA molecule*, which carries our inherited traits.

Organic, or inorganic – Does it *have* carbon, or *not?*

Quite amazingly, we can subdivide the entire family of molecules into only two major classes. These are the *organic molecules*, which include carbon atoms, versus the *inorganic molecules*, which do "not" (*-in*) contain any carbon:

Molecule 2

$$\begin{array}{ccc} \textbf{ALL BODY} = & \textbf{ORGANIC} & + & \textbf{INORGANIC} \\ \textbf{MOLECULES} & \textbf{(Contain C atoms)} & \textbf{(Do not contain C atoms)} \end{array}$$

The word, organic, literally "pertains to carbon" or "pertains to organs." This alternate identity clearly demonstrates that carbon atoms are the major building blocks out of which the body organs are constructed.

The Body's Inner Sea

The organic (carbon-containing) molecules play the chief role in providing the skeleton or scaffolding upon which the body cells and their organelles are built. But it is not only the cells themselves that are important. Just as critical for human survival are the *body fluids* that are located both inside of, and outside of, the cells.

THE BODY FLUIDS AS INORGANIC SOLUTIONS

The human body fluids can be generally classified as *inorganic saline* (**SAY**-leen) *solutions*. Saline "pertains to salt"; hence, an inorganic saline solution is a solution without many carbon atoms, but containing loads of salt. About 2/3 of Planet Earth is covered by an extremely huge saline solution – the salty water of all the oceans!

A solution begins with the action of a *solvent* (**SAHL**-vent) or chemical "dissolver." The solvent acts upon a particular *solute* (**SAHL**-yoot) or "thing dissolved":

SOLUTION = SOLVENT acting to dissolve a SOLUTE

The *extracellular* (**eks**-trah-**SELL**-you-lar) *fluid* lying "outside" (*extra-*) of our body's "little cells" (*cellul*), is one type of inorganic saline solution. The major solvent dissolver is the water molecule, H_2O. The chief solute in the extracellular fluid is *sodium chloride, NaCl.*

Sodium chloride (NaCl), or common table salt, occurs as a solid cube. Figure 4.2 shows how the + charges of Na^+ solute are attracted to the negative (−) charges on the O^{2-} of many H_2O molecules. Likewise, the negative (−) charges of Cl^- are attracted to the positive (+) charges of the H^+ poles or ends of the surrounding water molecules. Therefore, sodium and chloride split apart from each other and become individual Na^+ and Cl^- *ions* (**EYE**-ahns). An ion is simply an atom that has either an excess or deficiency of outermost electrons, so that it is electrically charged.

A

Electrolyte functions of ions

NaCl is a well-known *electrolyte* (ee-**LEK**-troh-**light**). An electrolyte is a substance that "breaks down" (*lyt*) into ions when placed into water solvent,

Atom 3

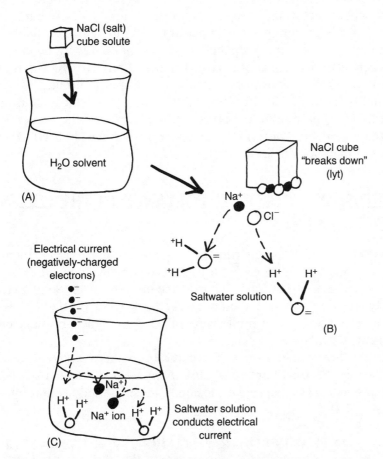

Fig. 4.2 A saline solution results from the action of H_2O upon NaCl.

such that the resulting solution can conduct an "electrical current" (*electro-*). The saline solution surrounding our body cells, then, is full of electrolytes and H_2O molecules with positively-charged H^+ poles. It is the body's inner sea. And, like the sea, it conducts an electrical current. This is because the negatively-charged electrons flowing in an electrical current are attracted to the many positively-charged areas within the extracellular fluid. Because of this seawater-like structural arrangement, you now know why your mother told you to get out of the water during a thunderstorm!

The Royal Carbon Quartet

Having considered the extracellular fluid as The Ocean Within, let us look more closely at the organic molecules inside of our bodies. We might speak about a Royal Carbon Quartet, the group of "four" (*quart*) major types of organic molecules that make up a large portion of *chemical anatomy*. We can define chemical anatomy as the part of human body structure that exists at the chemical level – subatomic particles, atoms, and molecules.

Beyond the saline (saltwater) body fluids, most of chemical anatomy consists of the organic molecules. The Carbon Quartet of these organic molecules are the *proteins*, *lipids* (**LIP**-ids), *carbohydrates* (**car**-boh-**HIGH**-drayts), and *nucleic* (new-**KLEE**-ik) *acids*.

PROTEINS: OF "FIRST" IMPORTANCE IN THE CELL'S ENVIRONMENT

Proteins get their Latin name because they are of "first" (*prot-*) importance. Specifically, the human body is largely glued together by a huge variety of *structural proteins*. Structural proteins make up most of the solid anatomy of the body. If you look very closely at the firm, solid portions of the skin, bones, and muscles, for example, you will find that their primary chemical component is structural protein.

There are three broad groups of structural proteins – *intracellular proteins*, found "within" (*intra-*) our "cells"; *membrane proteins*, as part of *cell membranes*; and *extracellular proteins*, located "outside" and around our "cells" (see Figure 4.3).

Molecule 3

STRUCTURAL = INTRACELLULAR + MEMBRANE + EXTRACELLULAR
PROTEINS   PROTEINS   PROTEINS   PROTEINS

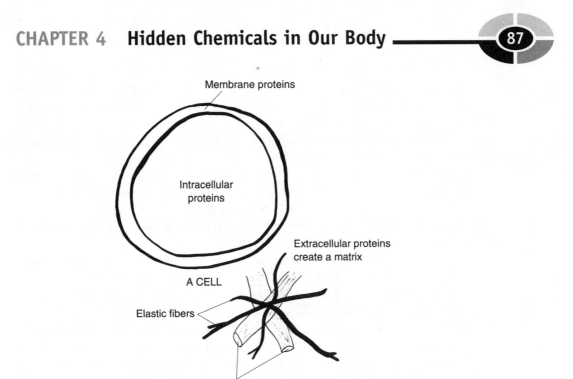

Fig. 4.3 The three broad types of structural proteins.

The intracellular proteins and membrane proteins will be covered in more detail in the chapter on cells (Chapter 5). Our focus, now, then, is upon the extracellular proteins.

Proteins and the extracellular matrix

We have already said that there is an extracellular fluid of saltwater surrounding most of our cells. For convenience, we will abbreviate the extracellular fluid as *ECF*. The ECF is not alone, however. Also abundant around our cells is an *extracellular matrix*. The extracellular matrix is a complex web or meshwork (matrix) of structural proteins located just "outside of our cells." It is therefore appropriate for us to describe a total *Extracellular Environment*. This Extracellular Environment consists of both the extracellular matrix of proteins, as well as the salty extracellular fluid (ECF) that circulates between these proteins.

| THE TOTAL EXTRACELLULAR ENVIRONMENT (Region around the cells) | = | THE EXTRACELLULAR MATRIX (Complex meshwork of structural proteins outside cells) | + | THE ECF (EXTRACELLULAR FLUID) |

A quick glance back at Figure 4.3 shows two prominent parts of the extracellular matrix, the *collagen* (**KAHL**-uh-**jen**) *fibers* and *elastic* (e-**LAS**-tik) *fibers*. Both collagen fibers and elastic fibers are *connective tissue fibers*. As their name suggests, connective tissue fibers are thin, *fiber*-like bundles of protein molecules that *connect* body parts together.

Collagen fibers consist of many thick, rope-like collagen molecules, which are literally "glue" (*colla*) "producers" (*gen*). Tough cuts of beef (actually cow skeletal muscle) have a high percentage of collagen fibers within their extracellular matrix. All of this collagen produces a glueing effect, such that these cuts of beef are very hard to chew.

Elastic fibers are rich in the highly elastic, rubbery protein called *elastin* (ih-**LAS**-tin). Elastin within the elastic fibers around blood vessels, for instance, allows them to stretch like a rubber band due to the "drive or push" (*elast*) of the blood pressure.

Of these two structural proteins, collagen is the much more abundant one in the extracellular matrix. Collagen molecules make up over 25% of all the protein found in the human organism! To better understand the importance of collagen, we will magnify a small part of a *tendon* (**TEN**-dun). A tendon is literally "a stretcher." A tendon is a thin, tough strap of *fibrous* (**FEYE**-brus) – "fiber" (*fibr*)-containing – *connective tissue* that attaches a skeletal muscle to a bone. The Latin name of tendon reflects its physiology: the tendon is stretched whenever the muscle contracts or shortens, thereby pulling upon the bone. As revealed in Figure 4.4, each tendon holds many long *collagen* (**CALL**-uh-jen) *fibers* running parallel to one another.

There are several different varieties of collagen molecules, but their basic structure is that of a *triple helix* (**HE**-licks) – three strands of linked *amino* (ah-**MEE**-noh) *acids*, "turned or rolled" (*helix*) together into a spiral shape. In general, an amino acid is a building block for the protein molecule. The *amino* part of the name indicates that the molecule begins with a nitrogen–hydrogen "amino" (NH_2) chemical group and the *acid* part shows that the molecule ends with an "acid" group called the *carboxyl* (car-**BAHK**-sul) or "carbon–oxygen" (COOH) *group*. The carboxyl group is called an "acid" because it tends to give off or donate its hydrogen atom as a positively-charged *hydrogen ion* (H^+). A protein such as collagen, therefore, consists of one or more long chains of bonded amino acids, which start with an NH_2 (amino) group and terminate with a COOH (acid or hydrogen-donating) group at the other end:

PROTEIN = A LONG CHAIN OF BONDED *AMINO ACIDS* (So starts with an *NH_2 amino* group, and ends with a *COOH acid* group)

Molecule 4

In the collagen molecule, there are 3 long chains of amino acids twisted tightly together to create a tough, stretch-resistant, triple spiral. This

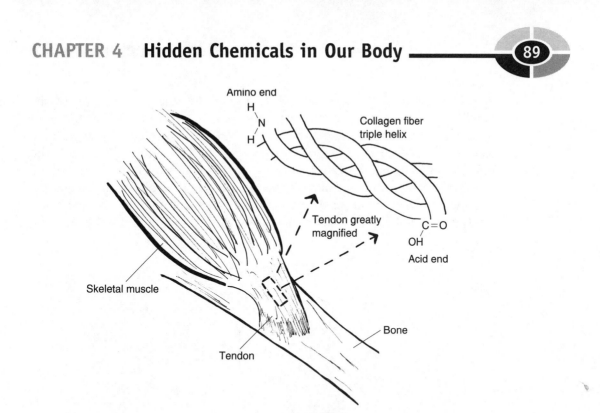

Fig. 4.4   Collagen: Main chemical structure of the tendon.

extremely high degree of Biological Order at the chemical level gives the collagen molecule an amazing *tensile* (**TEN**-sil) *strength* – powerful ability to resist pulling and tension forces. [**Study suggestion:** How does this characteristic of high tensile strength help the tendon do its job?]

LIPIDS: A VERY "FAT" CHEMICAL FAMILY!

The second member of the Royal Carbon Quartet comprise the lipids. The word root, *lip*, means "fat," while the suffix, -*id*, means "belonging to a group." [**Study suggestion: "If you're a relative of a lipid, you really need to go on a diet!"** Explain the reasoning behind this comment.]

The lipids are a group of organic molecules that contain many carbon–carbon (C–C) and carbon–hydrocarbon (C–H) bonds, so that they are *insoluble* (in-**SAHL**-yew-bl) or "not dissolvable" in water. There is an old chemical rule-of-thumb that, "Like dissolves like." This means that a solvent having electrically-charged molecules will tend to dissolve a solute with particles that are also electrically charged, because both the solvent and the solute are alike. Take the case of NaCl (sodium chloride) and water (H_2O). These two chemicals are very much alike in their electrical charge and chemical bonding. As we saw back in Figure 4.2, the Na^+ portion of the NaCl crystal is attracted to the net negative charge on the O^{2-} of the H_2O molecules. And the Cl^- portion of the NaCl is attracted to the positive charges on each of the H^+ poles or ends of the H_2O molecule. Hence, the particles of sodium chloride are very *soluble* (**SAHL**-yew-bl) or "dissolvable" in water, because both chemicals have areas of net (overall) electrical charge.

Now consider, in marked contrast, the chemical anatomy of the body lipids (Figure 4.5). The three main groups in the lipid family are the

Molecule 5

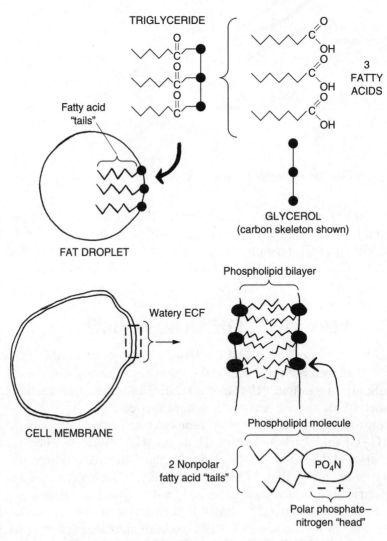

Fig. 4.5   The three major relatives in the lipid family.

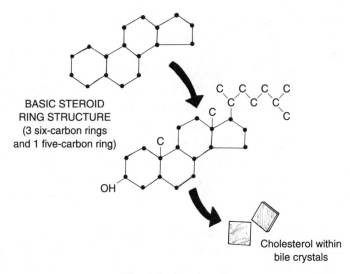

BASIC STEROID
RING STRUCTURE
(3 six-carbon rings
and 1 five-carbon ring)

OH

Cholesterol within
bile crystals

Fig. 4.5   (continued)

triglycerides (**try-GLIS**-er-eyeds), *phospholipids* (**fahs**-foh-**LIP**-ids), and *steroids* (**STEER**-oyds).

LIPIDS = TRIGLYCERIDES + PHOSPHOLIPIDS + STEROIDS

Triglycerides and phospholipids have an important part of their chemical anatomy in common: they both contain *fatty acid "tails"* that are strongly *hydrophobic* (**high**-droh-**FOH**-bik) or "water" (*hydr*) "hating" (*phobic*).

Each triglyceride molecule has "three" (*tri-*) fatty acid tails attached to a three-carbon molecule called *glycerol* (**GLIH**-ser-**ahl**). Triglycerides make up most of the body fat we humans store as extra energy within some of our cells and tissues. Hundreds of triglyceride molecules tend to group together and form sphere-shaped fat droplets.

Not surprisingly, the hydrophobic fatty acid tails are located deep within the fat droplet, far from any contact with water. But the glycerol end of each triglyceride molecule lies near the surface of the fat droplet, where it may come into contact with water. The fatty acid tails are almost entirely composed of bonded carbon–carbon and carbon–hydrogen atoms, which, having no net electrical charge, do not mix at all with charged H_2O!

The phospholipid molecules are an important anatomical component of most cell membranes. The phospholipids are arranged in two columns, with their hydrophobic fatty acid tails mixing together in the middle of the membrane, far from any saltwater. Since they are arranged in "two" (*bi-*) layers, they are often called the *phospholipid bilayer* (**BUY**-lay-er) of the membrane.

Each phospholipid molecule has a single *polar phosphate* (**FAHS**-fayt)–*nitrogen* head end, with both PO^{4-} (phosphate) and N^+ charged ends or "poles." The polar phosphate–nitrogen head, bearing positive and negative charges, is considered *hydrophilic* (**high**-droh-**FILL**-ik) or "water" (*hydr*) "loving" (*philic*). Thus, the charged phosphate head of the outer phospholipid layer sticks outward from the cell surface, contacting the watery extracellular fluid, which it so strongly "loves" (phil). Similarly, the charged phosphate heads of the inner phospholipid layer project inward to contact the watery intracellular fluid.

TRIGLYCERIDES   – **Contain** *Fatty Acid* =   **LIPIDS THAT**
AND PHOSPHOLIPIDS   **"Tails"**   **ARE ALSO** *FATS*

Finally, the last major group of lipids are the steroid molecules, which are literally "solid-oil" (*ster*) "resemblers" (*-oid*). They get this name involving "solid oil" due to the waxy appearance of their solid crystals, which often occur in oils. Since steroids do not contain fatty acids as part of their structures, they are not fats, at all! Rather, they are a group of non-fatty lipids that include closed rings of carbon atoms – three six-carbon rings, plus one five-carbon ring.

FATS = **LIPIDS THAT CONTAIN** *FATTY ACIDS*
= **TRIGLYCERIDES + PHOSPHOLIPIDS**
STEROIDS (Do *Not* **Contain** *Fatty Acids*) = **LIPIDS, BUT** *NOT FATS*

One important steroid is *cholesterol* (**koh**-**LES**-ter-ahl). Cholesterol is literally a "bile" (*chole*) "solid" (*ster*) that includes an "alcohol" (*-ol*) chemical group, -OH. As the *chole* portion of its name indicates, cholesterol is a very important solid, oily substance that occurs as square, scaly crystals within *bile* from the liver.

In addition to cholesterol, most of the so-called male and female "sex hormones" are also based upon the steroid ring structure.

CARBOHYDRATES – THE "CARBON–WATER" MOLECULES

Like the proteins and lipids, the carbohydrates have something important to say about themselves by looking at their general group name. The carbohydrates are literally "carbon" (*carbo*) "water" (*hydr*) molecules. This name reflects the fact that their *molecular* (**moh**-**LEK**-yew-lar) *formulas* can all be written as (CH_2O) multiplied by the same number; the molecules of the carbohydrates can all be represented *as if* they contained equal numbers of carbon atoms and water molecules. The glucose molecule, for instance, can have its molecular formula written as $(CH_2O)_6$, or multiplied out as $C_6H_{12}O_6$. (In reality, glucose and other carbohydrates do not contain actual water molecules – only their equivalent in the number and types of atoms found within their chemical structure.)

Within the human organism, the carbohydrate members of the Royal Carbon Quartet are mainly the *sugars* or *saccharides* (**SAK-ah-rides**), plus a large molecule called *glycogen* (**GLEYE-koh-jen**).

A

Molecule 6

MAIN CARBOHYDRATES =	SMALLER	+	GLYCOGEN
IN THE	SACCHARIDES		(A LARGE
HUMAN BODY	(THE SUGARS)		POLYSACCHARIDE)

The saccharides or sugars

The sugars are members of the saccharide group of carbohydrates that have a "sweet" taste. The *monosaccharides* (**mahn-uh-SACK-uh-rides**) or "single" (*mono-*) "sugars" are the simplest and smallest saccharides. Two very common monosaccharides are glucose and *fructose* (**FRUK-tohs**) or "fruit" (*fruct*) "sugar" (*-ose*). Glucose, of course, is the main carbohydrate within the bloodstream used as a fuel by the body cells. It is one of the final breakdown products of digestion of eaten carbohydrates (such as bread, potatoes, and other sugary or starchy foods). Fructose is found within raisins and other sweet fruit.

Glycogen as a polysaccharide

Individual monosaccharides (such as glucose and fructose) are sweet, simple sugars that are often broken down by the body and used directly for energy. These can be clearly contrasted with the *polysaccharides* (**pahl-e-SACK-uh-rides**), which contain "many" (*poly-*) "sugars" or smaller saccharides. Within the human body, the main polysaccharide is glycogen. Glycogen is literally a "producer" (*-gen*) of "sweetness or glucose" (*glyc*). Glycogen is stored in large quantities inside our muscle and liver cells. The glycogen molecule consists of a large number of glucose molecules, bonded together in long chains (Figure 4.6). Under conditions of exercise, fasting, or dieting, special *enzymes* (**EN-zighms**) – protein "fermenters or transformers" – tend to break the stored glycogen back down into individual glucose molecules. These free glucose molecules can then enter cells, and be used to ease any current energy shortage.

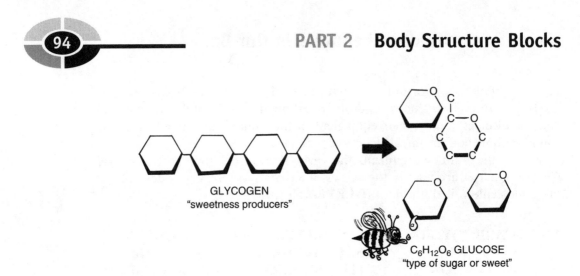

Fig. 4.6 Glycogen into glucose.

THE NUCLEIC ACIDS: DWELLERS WITHIN THE "KERNEL"

The final group among the Royal Carbon Quartet comprises the nucleic acids. The word, nucleic, literally "pertains to" (*-ic*) a "kernel, seed, or core" (*nucle*). But in this case, we do not mean a kernel or small rounded seed of corn. Rather, we are referring to the *cell nucleus*. You might recollect (Figure 4.1) that the atomic nucleus, a central "kernel" within each atom, contains protons and neutrons. Similarly, the cell nucleus is a rounded, kernel-like cell organelle that contains smaller structures (Figure 4.7, A).

The nucleus is surrounded by a thin *nuclear* (**NEW**-klee-ar) *membrane*. The nuclear membrane has pores or holes in it, through which certain chemicals can enter and leave the nucleus. Among these chemicals is one called *Messenger RNA*, abbreviated as *mRNA*. The letters, *NA*, you might have guessed, are for "*Nucleic Acid*." The *R* is an abbreviation for "*ribo-*." Hence, RNA is short for *ribonucleic* (**rye**-boh-new-**KLEE**-ik) *acid*. Ribonucleic acid is named for *ribose* (**RYE**-bohs), a "five-carbon sugar." Like the glucose molecule, ribose has its carbon atoms arranged into a ring shape. And on one of these carbons, there is an O (oxygen) atom attached (Figure 4.7, B).

There are a number of ribose subunits that occur along the length of the mRNA molecule, which takes the form of a single, relatively short strand of bonded atoms. Each messenger RNA molecule comes up alongside of a *chromosome* (**KROH**-moh-**sohm**), a dark-"colored" (*chrom*), wormlike "body" (*som*) within the nucleus.

Inside the chromosome (dark-colored body) is a *DNA molecule*. With *NA* again standing for "*Nucleic Acid*," the *D* in DNA is short for *deoxyribo-* (dee-**ahk**-see-**RYE**-boh). The complete abbreviation, DNA, thus stands for *deoxyribonucleic* (dee-**ahk**-see-**RYE**-boh-new-**KLEE**-ik) *acid*. Referring back

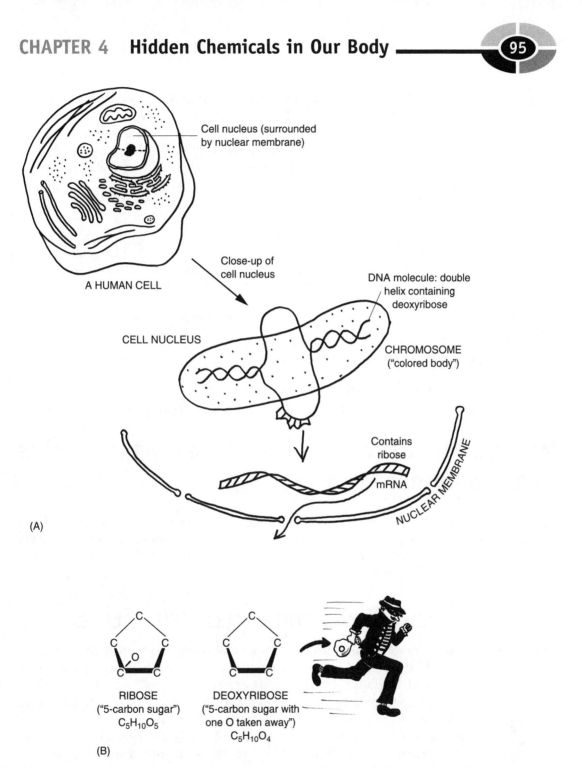

Cell nucleus (surrounded by nuclear membrane)

A HUMAN CELL

Close-up of cell nucleus

CELL NUCLEUS

DNA molecule: double helix containing deoxyribose

CHROMOSOME ("colored body")

Contains ribose

mRNA

NUCLEAR MEMBRANE

(A)

RIBOSE
("5-carbon sugar")
$C_5H_{10}O_5$

DEOXYRIBOSE
("5-carbon sugar with one O taken away")
$C_5H_{10}O_4$

(B)

Fig. 4.7 DNA and RNA: Nucleic acids found within the cell "kernel". (A) The cell nucleus and its contents. (B) Ribose versus deoxyribose.

to Figure 4.7 (B), we note that this chemical name clearly indicates that, "Our Poor Ribose Has Been Robbed!" You curtly reply, "Robbed of *what*?" The figure shows that the ribose has had its "oxygen" (*oxy*) atom taken "away from" (*de-*) it! So deoxyribose is a "ribose with one O taken away."

The DNA molecule, like what we saw earlier for the collagen protein molecule, is twisted into a spiral shape. But it takes the form of a double helix, rather than a triple helix. Deoxyribose sugar molecules occur at intervals along each twisted chain of this DNA double helix.

The DNA molecule and the mRNA (messenger RNA) molecules play critical roles in directing the process of *protein synthesis*. The details of this very essential process in the body are discussed in some depth within our companion volume, *PHYSIOLOGY DEMYSTIFIED*.

Nucleic acid summary

Overall, we can state a general word equation for the nucleic acids:

$$\text{NUCLEIC ACIDS} \quad = \quad \text{DNA} \quad + \quad \text{RNA}$$
(*Acids* within Cell *Nucleus*) (Contains *Deoxyribose*) (Contains *Ribose*)

Molecule 7

Mutations: Pathological Anatomy Often Begins with Abnormal Proteins

The structural proteins resulting from protein synthesis normally show an extremely high degree of Biological Order. For a specific example, we need only check back and admire the elegant twisted triple helix pattern of the collagen protein within a muscle tendon. (Review Figure 4.4, if desired.)

SCLERODERMA AND OTHER "COLLAGEN DISEASES"

Because collagen makes up over 25% of all our body's proteins, abnormal changes in the connective tissues involving collagen fibers can create severe pathological anatomy. Consider, for example, *scleroderma* (**sklir**-oh-**DER**-muh) – an abnormal "hardening" (*scler*) of the "skin" (*derm*). Scleroderma is a member of a group broadly called the *collagen diseases*.

In cases of scleroderma, part of the problem is an unusual swelling and fragmentation of the collagen fibers in the skin into disordered, smaller pieces. The skin becomes progressively thicker, stiffer, and harder, until movement of the fingers becomes nearly impossible! Because of this patho-

Molecule 1

logical anatomy at the molecule level of collagen, severe pathophysiology of the skin and other connective tissue-rich areas of the body may follow.

Quiz

Refer to the text in this chapter if necessary. A good score is at least 8 correct answers out of these 10 questions. The answers are listed in the back of this book.

1. An atom is best defined as:
   (a) A large piece of flesh visible at the skin surface
   (b) Several related structural compounds studied in chemistry
   (c) The smallest recognizable part of a chemical element
   (d) A short version of some particular molecule

2. The four most common elements in the human body are
   (a) Ne, O, C, P
   (b) Ca, N, O, H
   (c) Na, K, Cl, H
   (d) H, C, N, O

3. The Chemical Level exists:
   (a) Above molecules, but below cells
   (b) From subatomic particles up through molecules
   (c) Only in the World of the Atom
   (d) Below molecules, but above the subatomic level

4. The positively-charged particles within any atom are called:
   (a) Neutrons
   (b) High-energy electrons
   (c) Electron clouds
   (d) Protons

5. The main reason a molecule like H_2O usually doesn't totally fall apart is that it:
   (a) Combines with other molecules to create ordered filaments
   (b) The component atoms are linked together by chemical bonds
   (c) Electrons in orbit around one atom cannot be transferred to another atom
   (d) Nuclear protons, once activated, always rise to higher energy levels

6. Gasoline molecules coming out of a fuel pump and into your car are largely hydrocarbons, consisting of many:
   (a) NaCl crystals
   (b) Liberated ions
   (c) H=H double-bonded hydrogen atoms
   (d) C—H groups

7. Cubes of *sucrose* (**SOO**-krohs) or common table sugar quickly dissolve in hot tea. The water in the tea can thus be described as:
   (a) Solute
   (b) Organic solvent
   (c) Solution
   (d) Inorganic solvent

8. Structural proteins include:
   (a) Most enzymes and other "leaveners"
   (b) Many intracellular, extracellular, and membrane proteins
   (c) Those found within the cell membranes
   (d) Nothing of any consequence within the cells

9. The main monosaccharide in the bloodstream used for fuel by our cells:
   (a) Glucose
   (b) Glycogen
   (c) Polysaccharide
   (d) Ribose

10. They are lipids, but not true fats:
    (a) Cholesterol and other steroid molecules
    (b) Fatty acids
    (c) Phospholipids
    (d) Triglycerides

Body-Level Grids for Chapter 4

Several key body facts were tagged with numbered icons in the page margins of this chapter. Write a short summary of each of these key facts into a numbered cell or box within the *Body-Level Grid* that appears opposite.

Anatomy and *Biological Order* Fact Grids for Chapter 4:

SUBATOMIC PARTICLE
Level

1	2

ATOM
Level

1	2

ATOM
Level

3

MOLECULE
Level

1	2
3	4

5	6

7

Function and *Biological Disorder* **Fact Grids for Chapter 4:**

MOLECULE
Level

1

Anatomy of Cells and Tissues: Tiny Chambers Form the Body "Fabric"

After a lot of digging around in the dirt, carefully examining the organic and inorganic molecules of which we are ultimately composed – starting with the Hidden Chemicals in our Body Basement – it is now time to begin our steep climb. Yes, we must now begin our steep climb up the Great Body Pyramid, out of the silent darkness of atoms and molecules, and rise towards The Light.

The Cell Level – Where *Anatomy* Comes to *Life*!

The Light we rather poetically speak about is, of course, the Light of *Life*! And it is at the Cell Level of Body Organization that this Light of Human Life first begins to shine out of the chemical darkness.

But this was not really understood, way back in 1665, when an English *microscopist* (my-**KRAHS**-cope-ist) named *Robert Hooke* used a simple microscope to draw what he saw in a thin slice of dead tissue from a cork tree. Hooke was very impressed with the highly orderly pattern of rows and columns of hollow chambers in the cork (see Figure 5.1). He described these hollow chambers as "little boxes" or *cells*.

THE MODERN CELL THEORY

What Robert Hooke was viewing were dead, rigid-walled plant cells (in cork bark). Over a century later, two German scientists, *Schleiden & Schwann*,

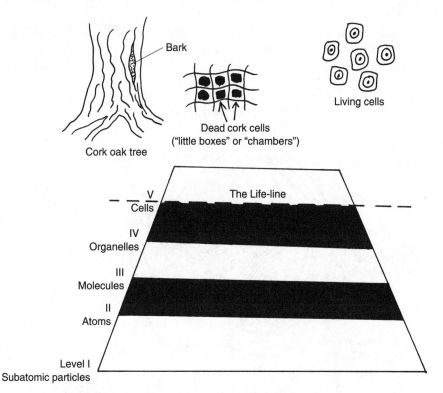

Fig. 5.1 The Cell Level and the "Body Basement" below it.

Cell 1

took a giant step forward in understanding by proposing the *Modern Cell Theory*. This theory clearly states that the cell is the basic unit of all living things. As such, the Cell Level (number V) is where the structures of the Great Body Pyramid begin to have true physiology, rather than merely body function. That is why we have earlier (Chapter 1) called the Cell Level the body's "Life-line." This means that it is at the level of size and complexity of the individual cell where body functions finally become sophisticated enough to create the conditions of something living.

ORGANELLES = "TINY ORGANS" WITHIN THE CELL

Yes, the *entire* cell *as a whole* is certainly alive. Therefore, we can best begin our study of the anatomy and physiology of the living human organism by looking at *cell anatomy* and *cell physiology*. But when we examine a human cell with a really powerful instrument, such as the electron microscope, something really interesting happens! After magnifying the cell thousands of times, an amazing variety of intermediate-sized structures appear in view: these structures are called the *cell organelles* (Figure 5.2).

Organelle 1

The cell organelles are literally the "tiny organs" present within each living cell. They are considerably bigger than the next smaller level – molecules. The cell organelles are *not* just bigger collections of molecules, however. Rather,

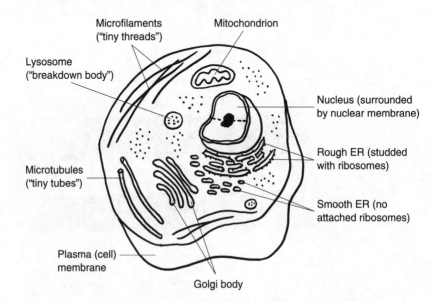

Fig. 5.2 An overview of the human cell and its major organelles.

they are tiny, organ-like structures that each carry out a certain highly specialized function within the cell.

"Okay," you may think, "If these tiny organelles are so organ-like, then how come *they* aren't alive, such as *real* organs, like the heart?"

We can answer this probing question with the *Little Organelle Rule-of-Thumb*:

INDIVIDUAL CELL ORGANELLES JUST *CAN'T "DO IT ALL"*! ORGANELLES ARE FAR *TOO SMALL* AND *TOO SPECIALIZED* IN THEIR STRUCTURES AND FUNCTIONS TO BE ALIVE!

Keeping an eye on Figure 5.2 for their pictures, let us discuss the micro-anatomy of the cell organelles, along with their specialized functions:

1. The cell or plasma membrane. The individual cell is surrounded by its own cell membrane or *plasma* (**PLAZ**-muh) *membrane*. The cell membrane is given the alternate name of plasma membrane because it is a thin "covering" (*membran*) "present" (*-e*) that helps "mold or form" (*plasm*) the shape of the cell. Thus, the cell (plasma) membrane helps explain why certain human cells are tall like columns, while others are short and squat!

Organelle 2

Another important function of the plasma membrane is its *selectively permeable* (**PER**-me-**ah**-bl) nature. To *permeate* (**PER**-me-**ayt**) something is to "pass through" it. Since the cell membrane is selectively permeable, this means that it lets some kinds of particles into and out of the cell while preventing the passage of others. Part of the reason for this selective permeability lies in the microscopic anatomy of the membrane.

Organelle 1

Earlier in this book (Chapter 4), we showed the cell membrane to be chiefly composed of a phospholipid bilayer – a double layer of phospholipid molecules having long fatty acid tails. We also mentioned the existence of membrane proteins. We classified these membranes as structural proteins. But there is yet another type within the membrane – *transport proteins*.

Figure 5.3 displays the characteristics of what has been called the *Fluid Mosaic* (moh-**ZAY**-ik) *Model* of the cell membrane. For our purposes, we will re-name it the "Jell-O Fruit-Salad Model"! Like clear, rubbery jello, the cell membrane is "fluid" in the sense of being relatively soft and able to shift its internal structure around. The heads of the phospholipid molecules might be pictured as little white marshmallows, bobbing up and down like the waves of the sea. The membrane is a "mosaic" in that it consists of a variety of different chemicals, not just phospholipids. The structural proteins may look like large pieces of sliced pineapple, very slightly rising with the waves of shaking Jell-O. Transport proteins with interior channels,

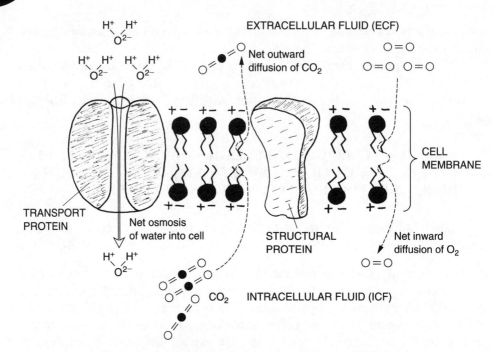

Fig. 5.3   The cell membrane: Fluid mosaic, or Jell-O salad?

used for carrying ions and other chemicals through the membrane, rather resemble cherries with holes made all the way through them with tooth-picks!

2. The nucleus. As the central "kernel" of the cell, the nucleus is sur-rounded by its own nuclear membrane. As explained in Chapter 4, the nucleus contains RNA and DNA, the two main types of nucleic acids. Depending upon the particular stage of division a cell is at, dark chromo-somes may also be visible within the nucleus.

Also visible within the nucleus is a smaller round *nucleolus* (new-**KLEE**-uh-lus) or "little kernel" (*nucleol*). The nucleolus contains a high concentra-tion of ribonucleic acid (RNA).

3. The cytoplasm. Slowly circulating outside of the nucleus is the *cytoplasm* (**SIGH**-toh-plazm) – the fluid "matter" (*plasm*) within the "cell" (*cyt*). One might not immediately recognize the cytoplasm as an organelle, because it is widely spread throughout the cell interior. True, the cytoplasm is not covered by its own membrane, unlike most of the organelles. But it does have a definite structure.

The cytoplasm contains an extensive *cytoskeleton* (**sigh**-toh-**SKEL**-eh-ton), around and between which slowly circulates the *intracellular* (**in**-trah-**SELL**-

yew-lar) *fluid*, abbreviated as *ICF*. The intracellular fluid is literally "within" (*intra-*) the cell.

THE CYTOPLASM =   CYTOSKELETON   +   ICF
(Rigid cell framework)   (Salty cell fluid)

Organelle 3

The cytoskeleton forms a rigid framework of support for the cell, somewhat like tentpoles holding up the flimsy canopy of a tent. The cytoskeleton in turn consists of two main types of protein rods – hollow *microtubules* (**my-kroh-TWO-byools**) and solid *microfilaments* (**my-kroh-FILL-ah-ments**):

CYTOSKELETON =   Microtubules   + Microfilaments
(Cell "skeleton"   ("Tiny" hollow   ("Tiny" solid
of protein rods)   "tubes")   "threads")

The protein rods of the cytoskeleton comprise a large part of the intracellular structural proteins mentioned in Chapter 4. The intracellular fluid or ICF circulating around the cytoskeleton (microtubules and microfilaments) contains lots of water, electrolytes (like Na^+ and Cl^- ions), and enzymes. These enzymes speed up many important chemical reactions. Thus, the cytoplasm has often been nicknamed the "factory area" of the cell.

4. The mitochondrion. If the cytoplasm is such a busy "factory," then it must use up a lot of *energy*! Energy is generally defined as the ability to do work. The kind of energy doing the work is called *free* or *kinetic* (kih-**NET**-ik) *energy*, because it is involved in "moving" (*kinet*) particles around. There is, of course, a huge amount of work performed within each living cell. The critical intracellular processes of membrane transport, protein synthesis, and cell division, for example, all require significant inputs of free (kinetic) energy.

The free or kinetic energy the cell uses chiefly comes from a molecule called *ATP*, which is an abbreviation for *adenosine* (ah-**DEN**-oh-seen) *triphosphate* (try-**FAHS**-fate). "So, where does this ATP (adenosine triphosphate) come from?" the inquiring brain may be prodded to ask.

Cell 1

"Most of it comes from the *mitochondria* (**my-toe-KAHN-dree-ah**)" is the correct answer to this question. Each *mitochondrion* (**my-toe-KAHN-dree-un**) is literally a "thread" (*mito*) "granule" (*chondr*) that is "present" (*-ion*) within a cell. As Figure 5.4 reveals, some mitochondria are long and slender (like a thread), while others are short and round (like a granule). This peculiar microanatomy explains their Latin name.

The internal anatomy of each mitochondrion is just as interesting. There is a double-membrane system: an *outer mitochondrial* (**my-toe-KAHN-dree-al**) *membrane* surrounds the organelle, while an *inner mitochondrial membrane* is

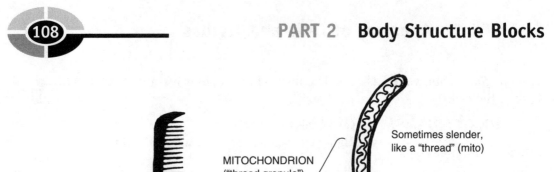

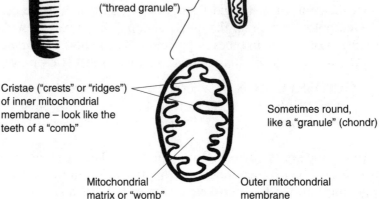

Fig. 5.4 The internal anatomy of the mitochondrion.

folded into structures called *cristae* (**KRIS**-tee). The cristae are slender "crests" or "ridges" that project into the *mitochondrial matrix* – the inner hollow "womb"-like cavity that occupies the center of each mitochondrion.

Both the cristae and the matrix of the mitochondrion work together to produce plenty of ATP during *aerobic* (air-**OH**-bik) *glycolysis* (gleye-**KAHL**-uh-sis). This is literally the "air" (*aer*) or oxygen-using process of "breaking down" (*lysis*) "glucose" (*glyc*). Because of its great ability to efficiently produce many ATP molecules from a single molecule of glucose, the mitochondrion has frequently been described as the "powerhouse" of the cell.

5. The lysosome. Instead of breaking down glucose to produce ATP, there is another organelle, called the *lysosome* (**LIE**-soh-**zohm**) or "break-down" (*lys*) "body" (*som*), that seems to *lyse* (**LICE**) or break down just about everything else! The lysosome is a spherical body surrounded by a membrane. It is important for the many body cells that engage in *phagocytosis* (**fay**-go-sigh-**TOH**-sis) – the "process of" (-*osis*) "cell" (*cyt*) "eating" (*phag*). A human white blood cell, for example, may *phagocytose* (**fay**-goh-**SIGH**-tohs) an invading *bacterial* (back-**TEER**-e-al) *cell*. After engulfing the bacterial cell, many of the lysosomes in the white blood cell rupture and release *digestive enzymes*. These enzymes speed the breakdown of the eaten bacterial cell into its component parts, which are eventually used by the white blood cell for its own ATP production.

"Why doesn't the human body become choked with millions of its *own* dead or dying cells?" a reader might ask in follow-up. The credit for avoiding such congestion with dead cells largely belongs to the process called *cell autolysis* (aw-**TAH**-lih-**sis**). Autolysis is the automatic "self" (*auto-*) "breakdown" (*lys*) of a dead or dying cell. This self-breakdown occurs due to the rupture of many of the dying cell's lysosomes, all at once. Thus, the massive flood of digestive enzymes released from many lysosomes destroys the dying cell from within, essentially exploding it out of existence! (Think of the dying cell as a self-exploding hand grenade!)

6. The endoplasmic reticulum. We have seen that some cell organelles are involved in *de*struction (as with the lysosome's digestive enzymes), while others are important for *con*struction. "What organelle besides the nucleus is involved in synthesizing proteins?" is one of the pivotal questions regarding cell construction. The answer is the *ER*, which is short for the *endoplasmic* (**en**-doh-**PLAZ**-mik) *reticulum* (**reh**-**TIK**-yoo-lum). The endoplasmic reticulum is "a tiny network" (*reticulum*) of flattened sacs "within" (*endo-*) the "cytoplasm" (*plasm*). The primary functions of the ER become obvious when you consider its two main types. The *rough ER* (rough endoplasmic reticulum) has its flattened sacs studded with many dark *ribosomes* (**RYE**-boh-**sohms**). The ribosomes are dark round "bodies" (*-somes*) containing "ribose" (*ribo-*). If you recall (Chapter 4), ribose is part of RNA, which is critical for protein synthesis.

Some of the cell's ribosomes are scattered throughout its cytoplasm, while others dot the surface of the rough endoplasmic reticulum. [**Study suggestion:** Imagine running your fingers over the surface of the ER, and feeling its "rough" surface of tiny ribosome bumps.] Because of its ribosomes, the rough ER is a staging platform within the cell where proteins are manufactured.

The second type of ER is the *smooth endoplasmic reticulum*. Obviously, the *smooth ER* is a tiny network of flattened sacs that do *not* have any ribosomes attached to them (i.e. running your fingers over the smooth ER would yield a non-bumpy feeling). The smooth ER, then, is not involved in protein synthesis; rather, it serves as an important intracellular transport system. It can be thought of as a busy highway system of hollow sacs and tubules carrying manufactured chemicals from one place to another within the cell.

In summary,

A

ROUGH ER = A "BUMPY" NETWORK OF FLATTENED SACS STUDDED WITH RIBOSOMES INVOLVED IN PROTEIN SYNTHESIS

Organelle 4

SMOOTH ER = A "NON-BUMPY" NETWORK OF FLATTENED SACS THAT CIRCULATE MATERIAL AROUND IN THE CELL

7. The Golgi body. *Camillo Golgi* (**GAHL**-jee) was an Italian *histologist* (hiss-**TAHL**-oh-**jist**) – "one who specializes in the study of" (*-ologist*) various kinds of "tissue" (*hist*). Most histologists, of course, are also pretty good *cytologists* (**SIGH**-tahl-oh-**jists**) or "specialized cell studiers," as well. Hence Camillo Golgi, peering at tissue cells through his rather primitive early microscope, first described the *Golgi* (**GAHL**-jee) *body or apparatus*, which now forever bears his name.

The Golgi body (Golgi apparatus) is a pancake-like stack of flattened sacs that seem to do the same thing – package the proteins, lipids, hormones, and other types of molecules that are constructed by the cell.

The Centrosome and Its Centrioles: Curly Star-shaped Cylinders for Mitosis

We have now introduced and discussed seven major types of cell organelles (including the cytoplasm, itself, within this organelle group). Another intracellular structure that could have been added is called the *centrosome* (**SEN**-troh-**sohm**). This is a "centrally" (*centr*)-located "body" (*som*) or region of the cytoplasm near the cell nucleus. (Study Figure 5.5.)

Rather than being a membrane-covered body (like a mitochondrion or a lysosome), however, the centrosome seems to be a *microtubule organizing center* within the middle of the cell. Focusing in upon the centrosome area with a *light* microscope (using light rays), the cytologist/histologist sees that it mainly consists of two *centrioles* (**SEN**-tree-**ohls**) or "tiny" (*-oles*) round "centers" (*centr*). Now, employing a really highly magnifying *electron* microscope (focusing beams of tiny electrons through the cell), each of these centrioles is seen to hold nine sets of "triple-packaged" microtubules.

Here we have another amazing example of an extremely high degree of Geometric Order within Human Bodyspace, this time made evident at the level of the organelles! The triple-packaged microtubules are arranged into a beautiful hollow cylinder that looks like a curly star or pinwheel when viewed end-on. The sets of microtubules making this curly star pattern are held together by smaller microtubules.

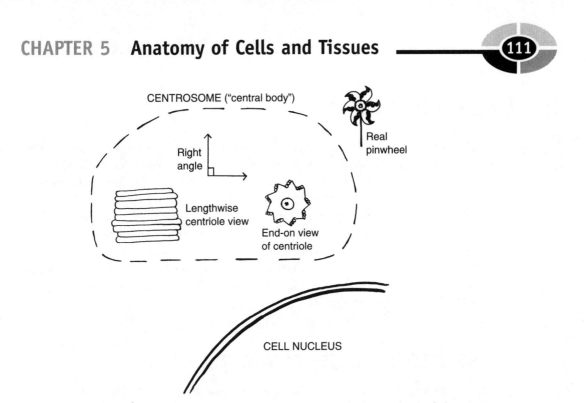

Fig. 5.5   The centrosome and its centrioles: High geometric order at the level of the organelles. Each centriole is a curly pinwheel cylinder of nine triple-packed microtubules.

Further accenting the striking phenomenon of High Geometric Order within the World of the Cell, there are usually two centrioles within the same centrosome. These two centrioles (each a curly pinwheel cylinder of nine triple-packed microtubules) are oriented at a right angle to each other. [**Study suggestion:** Go way back and just take a quick look at Chapter 2. What was said, there, about the right angle, rectangle, and the origin of the idea of anatomic planes? – So, weren't the Ancient Egyptians really *on* to something very fundamental and basic about the underlying Geometry of Nature, even at the Level of the Cell Organelles?]

In summary,

A

Organelle 5

THE CENTROSOME = A CENTRAL CELL REGION OR BODY THAT ORGANIZES MICROTUBULES, CONSISTING OF TWO CENTRIOLES

And,

EACH CENTRIOLE = A CURLY, STAR-SHAPED HOLLOW CYLINDER OF 9 TRIPLE-PACKED MICROTUBULES

The Cell Cycle and Mitosis

The stage is now set to consider the process of *mitosis* (my-**TOH**-sis). The centrosome and its two curly star-shaped cylinder centrioles are right there, in the center of the cell. They help set the stage for the *Cell Cycle* and its "condition of" (*-osis*) chromosome "threads" (*mit*), that we technically call mitosis.

What do we specifically mean by the phrase, Cell Cycle? The Cell Cycle is the entire life span of a particular cell. It starts when the cell is produced from its previous *parent cell*, and ends with the orderly subdivision of both the cell nucleus and its cytoplasm into two new *daughter cells*. Hence, mitosis is critically involved in the Cell Cycle as the dividing process involving the nucleus and its chromosomes.

THE SPECIFIC PHASES OF THE CELL CYCLE

1. Interphase. When the cell is "between" (*inter-*) actively dividing, it is said to be in the state called *interphase* (**IN**-ter-**fayz**). During interphase, the DNA molecules within the cell nucleus start out as thin, dark strands of *chromatin* (kroh-**MAT**-in) – "colored" (*chromat*) DNA material covered with "protein" (*-in*) (see Figure 5.6). The DNA within the chromatin threads duplicates – making a copy of itself.

The two centrioles of the centrosome also make duplicate copies of themselves. Thus, there are now two centrosomes, containing a total of four centrioles, within the parent cell.

The nucleolus is still visible within the larger cell nucleus. This indicates that protein synthesis and cell growth are continuing throughout interphase.

2. Prophase. Following interphase, comes *prophase* (PROH-fayz). It is the "first" (*pro-*) phase of mitosis. Prophase begins as soon as the duplicated DNA within the chromatin threads has coiled up and condensed into individual chromosomes. Since the DNA has already made a copy of itself during interphase, there are now 46 pairs of *sister chromatids* – original chromosomes plus their duplicates – held together in the middle by a *centromere* (**SEN**-troh-**meer**) or "central segment."

The nuclear membrane starts to disappear. A *mitotic* (my-**TAH**-tik) *spindle* is created. The mitotic spindle is named for its resemblance to an old-fashioned sewing spindle, being wider in the middle, and tapering towards both ends. The spindle is made by the orderly arrangement of microtubules, which extend from the centrioles and eventually push them out to opposite poles of the cell.

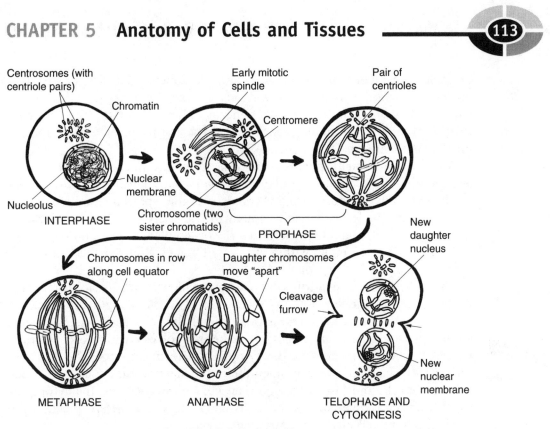

Fig. 5.6   The Cell Cycle and its stages.

Not being held in by the nuclear membrane anymore, the sister chromatids are contacted by microtubules, which move them towards the middle of the cell.

3. Metaphase. As its name indicates, *metaphase* (**MET**-ah-**fayz**) is the "phase" occurring "after" (*meta-*) prophase. By this stage, the mitotic spindle is completely formed, with a pair of centrioles at either pole or end. The sister chromatids are pushed into a single horizontal row along the *equator* or middle of the cell.

4. Anaphase. Up to this time, each of the 46 duplicated chromosome pairs or sister chromatids have all stayed attached together. In *anaphase* (**AN**-uh-**fayz**), however, the centromeres split, and each of the duplicated chromatid sisters finally move "apart" (*ana-*) from one another. The 92 separated, new chromosomes (often called the *daughter chromosomes*) are now pulled towards opposite poles of the cell, by their attached microtubules.

5. Telophase and cytokinesis. The "phase" that formally "ends" (*telo-*) mitosis is technically called *telophase* (**TELL**-uh-**fayz**). In telophase, many of the events that happened during prophase are exactly reversed! For example, remember that in prophase the nuclear membrane disappeared. But during telophase, two new *daughter nuclei*, each with its own individual *nuclear membrane*, start to appear, one at either end of the dividing cell. Further, the mitotic spindle (which first arose during prophase) eventually breaks apart into tiny fragments.

All of the four main phases of mitosis (prophase, metaphase, anaphase, and telophase) essentially describe a *division of the original cell's nucleus and its contents*. But just duplicating chromosomes and splitting the nucleus into new daughter nuclei really isn't enough to end the Cell Cycle, is it? [**Study suggestion:** What would happen if, say, the original parent cell still remaining at telophase was left intact within our body tissues? Would such a cell still be considered normal? Why, or why not?]

Enter *cytokinesis* (**sigh**-toh-kih-**NEE**-sis), which is a "division or movement" (*kines*) of the "cytoplasm" (*cyto-*) into two parts. Cytokinesis occurs at the same time that telophase is proceeding. But instead of focusing upon the nucleus and its chromosomes, cytokinesis involves creation of a *cleavage furrow*, and what happens to the cytoplasm after it appears. The cleavage furrow is basically a furrow or indentation that starts at the equator on either side of the cell, and just keeps going deeper and deeper into the cytoplasm from each side. Eventually, the cleavage furrow completely pinches the original parent cell into two new *daughter cells*.

Telophase (and the rest of mitosis) have already finished by this complete division of the original cell's cytoplasm. Therefore, the two new daughter cells pinched off from the original parent by cytokinesis have a full package of all 46 chromosomes, contained within their own nucleus and nuclear membrane. The two daughter cells each go into an interphase of their own, and the seemingly endless magic of the Cell Cycle begins anew.

To capsulize:

Cell 2

THE CELL CYCLE = INTERPHASE   + MITOSIS   + CYTOKINESIS
                                        (Division of   (Division of
                                        the Nucleus)   the Cytoplasm)

INTERPHASE   = The Phase "between" One Cell Division and Another

MITOSIS   = PROPHASE + METAPHASE + ANAPHASE
                      + TELOPHASE
                    = All Processes Involved in *Division of the*
                      ***Nucleus* of the Original Parent Cell**

CYTOKINESIS   = All Processes Involved in *Division of the Cytoplasm*
                      of the Original Parent Cell

Tissues: Families of Related Cells

We have completed our basic survey of the Chemical Level (I–III) and the Organelle and Cell Levels (IV and V) of the Great Body Pyramid. Level VI, the Tissues and their anatomy, will now occupy our attention.

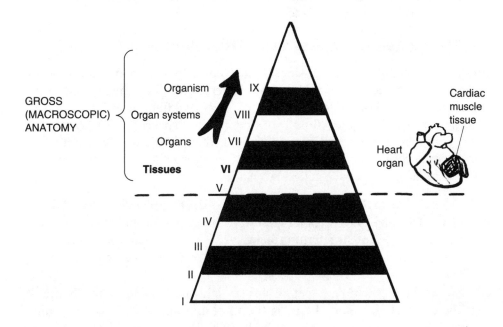

TISSUES AS THE BODY "FABRIC"

Way back in Chapter 1, we made mention of Andreas Vesalius (Father of Anatomy), who wrote a famous book, *On The Fabric of The Human Body*. And in Figure 1.5, we followed his lead and compared all body structures to a "woven fabric" having a grid or matrix pattern. For practical purposes, many anatomists and artists of the human form have utilized the woven fabric or grid idea as their geometric model for visually representing the body.

The word, *tissue*, actually derives from the Old French and Latin for "(something) woven." Therefore, having tissues represent the idea of Human Bodyspace as a "woven fabric" is quite appropriate! To be specific, a tissue is defined as a collection of similar cells plus the *intercellular* (**in**-ter-**SELL**-you-lar) *material* located "between" (*inter-*) them.

Tissue 1

TISSUE = A COLLECTION OF + THE *INTER*CELLULAR MATERIAL
**   SIMILAR CELLS   LOCATED *BETWEEN* THE CELLS**

The solid portion of the intercellular material is a part of the extracellular matrix of structural proteins located outside cells (Chapter 4). This complex meshwork of structural proteins does make tissues in general physically resemble a grid or "woven fabric."

The related concept of interstitial fluid

But the intercellular material doesn't just consist of a solid meshwork of proteins between cells. It also includes a watery, fluid component – the *interstitial* (in-ter-**STISH**-al) *fluid*. The phrase, interstitial fluid, translates to mean "pertaining to" (*-ial*) the "fluid" located in the "spaces" (*stit*) "between" (*inter-*) the tissue cells. The interstitial fluid, then, is the fluid circulating in the spaces between the solid meshwork of criss-crossing structural proteins found outside of our tissue cells.

THE   = THE EXTRACELLULAR + INTERSTITIAL FLUID
INTERCELLULAR   MATRIX (Solid meshwork   (Fluid in spaces between
MATERIAL   of structural proteins   the mesh of proteins
**   outside cells)   outside tissue cells)**

Now, let's take the time to get all this straight! Remember from Chapter 4 that the extracellular fluid (ECF) is the total amount of salty water (saline solution) located outside of *all* our body cells. We are now, therefore, clarifying the notion of the ECF by subdividing it into two major types of fluids – the interstitial fluid plus the *plasma* (**PLAZ**-mah). Plasma is the watery portion of the ECF present between the individual blood cells, within the bloodstream (see Figure 5.7).

In overview:

THE TOTAL   =   INTERSTITIAL FLUID   +   BLOOD PLASMA
EXTRACELLULAR   (Portion of ECF located within   (Portion of ECF found
FLUID OR ECF (All of the   spaces of the extracellular   between the cells of
fluid outside the body cells)   matrix of proteins)   the bloodstream)

The Four Basic or Primary Tissues

The tissue level of body organization is best represented by the four *basic* or *primary tissues*. These are the four *general* types of tissue which have the

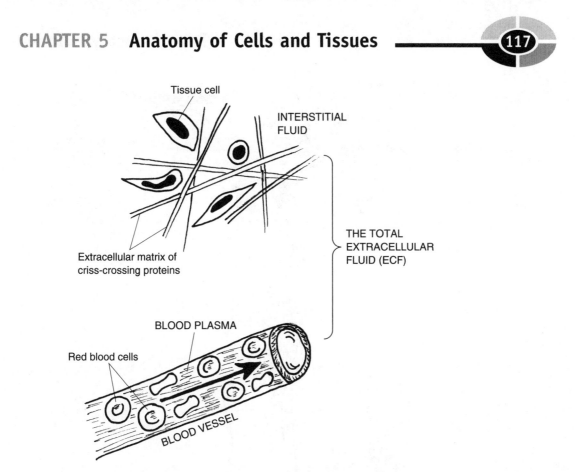

Fig. 5.7   The two main parts of the extracellular fluid.

common characteristics into which all of the *specific* types of body tissue can be grouped.

For ease of memorization, we are going to give you a *CNEMI* (**NEE**-mee), or "leg" up! And this CNEMI (helping leg) is actually shown in Figure 5.8. Besides being a Latin word root for "leg," we are using CNEMI as a helpful abbreviation for the basic (primary) tissues, plus the skin.

C IS FOR CONNECTIVE TISSUE

The first letter in *CNEMI* is for *Connective Tissue*. As its name advertises, connective tissue either directly or indirectly *connects* body parts together. Perhaps the most representative (and most common) type of connective tissue is called *areolar* (ah-**REE**-uh-**lar**) or *loose connective tissue*. Areolar (loose) connective tissue is mainly a criss-crossing meshwork of thick

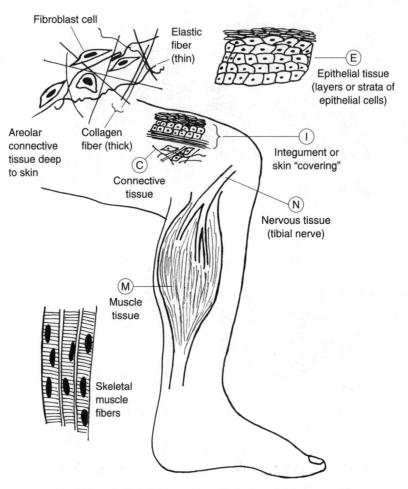

Fig. 5.8 The helpful *CNEMI* – Basic tissues of the "leg."

collagen fibers, and many thin elastic fibers, loosely woven together. (These fibers were described back in Chapter 4.) The word, areolar, "refers to" (*-ar*) "little areas" (*areol*). Because of their extensive crossing-over, the mesh-work of collagen and elastic fibers has many *areoli* (ah-**REE**-oh-**lie**) or "little areas" filled with interstitial fluid between them. Hence, areolar connective tissue is one of the best models available to literally describe Human Bodyspace as a *real* grid or matrix!

There are numerous *fibroblasts* (**FEYE**-broh-**blasts**) or "fiber" (*fibr*) "for-mer" (*-blast*) cells visible within any prepared slide of areolar connective tissue. This is because the fibroblasts are the cells that secrete or produce the extensive meshwork of collagen and elastic fibers.

Areolar connective tissue forms a loose packing deep to the skin, between the skin and skeletal muscles, and almost everywhere else in the body where there is some space available. This does, of course, include the empty spaces within the actual *CNEMI* or ("leg")!

N IS FOR NERVOUS TISSUE

The second letter in *CNEMI* is for *Nervous Tissue*. Nervous tissue does "pertain to" (*-ous*) the "nerves" (*neur*) – the slender white cords that carry information from one part of the body to another. The nerves are long extensions of various neurons or nerve cells. Consider, for instance, the *tibial* (**TIB**-ee-al) *nerve*, which runs right down along the *tibia* (**TIB**-ee-ah) or shin-bone in the lower leg.

E IS FOR EPITHELIAL TISSUE

The third letter in *CNEMI* is for *Epithelial* (**eh**-pih-**THEE**-lee-al) *Tissue*. The epithelial tissue literally "pertains to" (*-al*) something present "upon" (*epi-*) the "nipples" (*theli*). This seems rather odd, but the epithelial tissue in general is a covering and lining tissue. It covers the body surface (including the nipples and the legs) and lines the interior of the body cavities. Epithelial tissue is also unique among the tissues in that it almost entirely consists of tightly-packed epithelial cells, with little or no intercellular material between them.

M IS FOR MUSCLE TISSUE

The fourth letter in *CNEMI* is for *Muscle Tissue*. The muscle tissue consists mainly of dozens of slender, rod-shaped *muscle fibers*, running parallel to one another. In some of the skeletal muscles attached to the bones of the leg, for instance, the muscle fibers are excited by the tibial nerve, which causes them to shorten or contract. Movement of the leg thus occurs.

I IS FOR THE INTEGUMENT

The fifth and final letter in *CNEMI* is for the *Integument* (in-**TEG**-you-ment). As noted in Chapter 4, the integument is our body "covering." "But, Professor Joe! I thought you just told us that epithelial tissue was the covering and lining tissue!" the alert reader may protest. Yes, but epithelial *tissue*, does, indeed, form the outer portion of the skin or integument of the leg, which is really a flat, membrane-like *organ*!

Organs = Specialized Collections of the Basic Body Tissues

Having introduced the basic or primary types of tissues, it is now appropriate to climb up one more horizontal layer in the Great Body Pyramid. This layer is Level VII – the *Organ Level*.

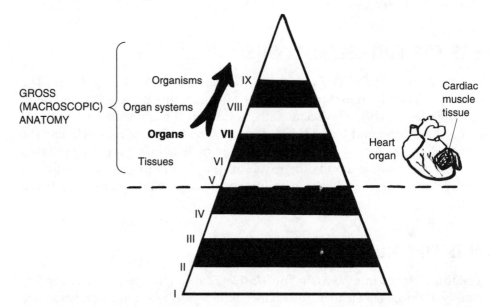

DEFINITION OF AN ORGAN

Organ 1

An *organ* is a collection of two or more of the basic body tissues, which together perform some specific body function. We have just said, for example, that the skin or integument is an organ, rather than a tissue. The skin contains all four of the basic or primary tissues. Its outermost layer is epithelial tissue. Its middle layer is mostly connective tissue. Nervous tissue with nerves and sensory receptors, densely supply the skin. And there is just a little bit of muscle tissue around the blood vessels in the skin.

Taken altogether, these four basic tissue types help make the skin our chief organ of protection and sensation.

Why Cancer Is so "Crabby": The Principle of Cellular Pathology

Early in this chapter, we explained the Modern Cell Theory, which holds that the cell is the fundamental unit of life. But what about its mirror image? If the normal, healthy cell is where true *physiology* begins, then is it not also at the cell level where disturbances in body structure and function begin to show up as true *pathophysiology*?

THE PRINCIPLE OF CELLULAR PATHOLOGY

Back in Chapter 4, we noted how pathological anatomy often begins at the Chemical Level, within the molecules, as in mutations causing abnormal proteins. One of the first places such abnormal changes or mutations in proteins are likely to show up is the next higher level on the Body Pyramid – the Cell Level.

The Principle of Cellular Pathology was first introduced by *Rudolf Virchow* (**FEAR**-koh), a German pathologist, in about the year 1856. It was Virchow who emphasized that the cell is the fundamental unit in pathology. Therefore, it is the pathological anatomy within the cell that often leads to the pathophysiology observed in various diseases.

A

Cell 1

Rudolf Virchow, for instance, was the first to describe *leukemia* (loo-**KEY**-me-ah) as "an abnormal condition of" (*-ia*) "white" (*leuk*) "blood" (*em*). Leukemia is considered a deadly type of *cancer*, which comes from the Latin for "crab."

The major problem with cancers in general seems to be the accumulation of errors in cell DNA. When there are too many errors in cell DNA, the phases of mitosis may become highly disordered. As a result, too many highly abnormal cells are produced. These abnormal cells multiply totally out of control, until their great degree of structural disorder results in major disturbances in physiology of the tissues containing them.

Dr. Virchow coined the term *leukocytosis* (**loo**-koh-sigh-**TOH**-sis) – an "abnormal condition of" (*-osis*) too many "white" (*leuk*) "cells" (*cyt*) within the bloodstream. The exact cause of leukemia remains unknown, but one of the major problems is an extensive leukocytosis. There are simply too many highly abnormal *leukocytes* (**LOO**-koh-**sights**) or "white blood cells" circulating within the bloodstream.

The normal range for the *total white blood cell count (WBC count)* is from 4,000 WBCs up to 10,500 WBCs per *microliter* (**MY**-kroh-**lee**-ter) or *cubic millimeter* of blood (see Figure 5.9). During leukemia, however, the uncontrolled mitosis of *stem cells* within the *red bone marrow* creates a severe leukocytosis (more than 10,500 WBCs per microliter of blood). There may be as many as 1,000,000 (one million) immature, highly abnormal leukocytes present in a single microliter of blood! No wonder that Dr. Virchow called the condition leukemia or "white blood," because he saw so many white blood cells in his microscope when he looked at the blood of a stricken patient!

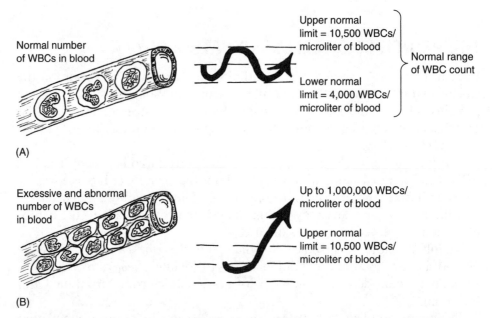

(A)

(B)

Fig. 5.9   Severe leukemia and the WBC (white blood cell) count. (A) A normal WBC count. (B) Severe leukocytosis (as in leukemia).

Actually, leukemia is more accurately described as a cancer of the red bone marrow, where most of the leukocytes are produced. Abnormal bone marrow cells, and lack of clotting cells in the bloodstream, may result in such fatal conditions as severe *hemorrhage* (**HEM**-or-**ahj**), or "bleeding," from which the afflicted patient does not recover.

Quiz

Refer to the text in this chapter if necessary. A good score is at least 8 correct answers out of these 10 questions. The answers are listed in the back of this book.

1. The English microscopist who first described cells under the microscope:
   (a) Louis Pasteur
   (b) Robert Hooke
   (c) Emperor Maximillian
   (d) Sir Robin of Loxley

2. The cell is the basic unit of all life:
   (a) The Least Squares hypothesis
   (b) "An apple a day keeps the doctor away!"
   (c) The Modern Cell Theory
   (d) The Principle of Cellular Pathology

3. "Tiny organs" present within each cell:
   (a) Organelles
   (b) Molecules
   (c) Tissues
   (d) Cellular macrostructures

4. Forms a selectively permeable barrier around the cell:
   (a) Mitochondrion
   (b) Golgi body
   (c) Lysosome
   (d) Plasma membrane

5. The Cytoplasm = ____ + ____:
   (a) Nucleus; nucleolus
   (b) Ribosomes; mitochondria
   (c) Cytoskeleton; ICF
   (d) Extracellular matrix; nucleus

6. The aerobic "powerhouse" of the cell:
   (a) Rough ER
   (b) Mitochondrion
   (c) Phospholipid bilayer
   (d) Cytoplasm

7. Has its surface studded with ribosomes:
   (a) Smooth endoplasmic reticulum
   (b) Nucleolus
   (c) Lysosome
   (d) Rough endoplasmic reticulum

8. The centrosome's chief function:
   (a) Controller of protein synthesis within cells
   (b) Activates breakdown of glycogen
   (c) Microtubule organizing center within the cell
   (d) Connecting joint for all organelles

9. The phase between the active cell dividing process is:
   (a) Prophase
   (b) Anaphase
   (c) Metaphase
   (d) Interphase

10. The four basic or primary types of tissue:
    (a) Connective, nervous, muscle, and membranous
    (b) Epithelial, serous, Golgi body, and lysosomal
    (c) Connective, nervous, epithelial, and muscle
    (d) Blood, guts, boils, and trouble!

Body-Level Grids for Chapter 5

Several key body facts were tagged with numbered icons in the page margins of this chapter. Write a short summary of each of these key facts into a numbered cell or box within the appropriate *Body-Level Grid* that appears below.

Anatomy and *Biological Order* Fact Grids for Chapter 5:

A

ORGANELLE
Level

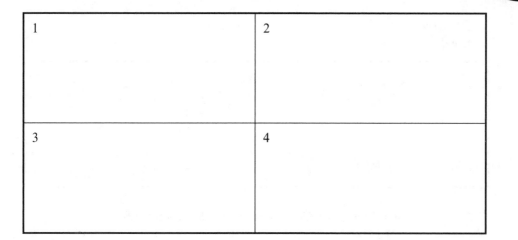

1	2
3	4

5

CELL
Level

1

TISSUE
Level

1

ORGAN
Level

1

Physiology and *Biological Order* Fact Grids for Chapter 5:

CELL
Level

1	2

Function and *Biological Order* Fact Grid for Chapter 5:

ORGANELLE
Level

1

Anatomy and *Biological Disorder* Fact Grid for Chapter 5:

A

CELL
Level

1

Test: Part 2

DO NOT REFER TO THE TEXT WHEN TAKING THIS TEST. A good score is at least 18 (out of 25 questions) correct. Answers are in the back of the book. It's best to have a friend check your score the first time, so you won't memorize the answers if you want to take the test again.

1. The basic building block element of organic chemistry is:
   - (a) Hydrogen
   - (b) Oxygen
   - (c) Potassium
   - (d) Carbon
   - (e) Nitrogen

2. **SUBATOMIC PARTICLES** = ___ + ___ + ___:
   - (a) Neutrons; Electrons; Protons
   - (b) Molecules; Electrons; Atoms
   - (c) Atoms; Molecules; Protons
   - (d) Nuclei; Electrons; Neutrons
   - (e) Molecules; Organelles; Protons

3. Chemical bonds always involve:
   (a) Total loss of all electrons from the bonded atoms
   (b) Creation of C−H linkages
   (c) Linkages between the outer electron clouds of two or more atoms
   (d) Exactly equal sharing of outermost electrons
   (e) Complete destruction of all subatomic particles involved

4. The saline solution surrounding our body cells is capable of conducting an electrical current because:
   (a) Its H_2O molecules are just "watery electrons" to begin with!
   (b) The contained solutes are electrolytes that mix with charged water solvent
   (c) There aren't any nonelectrolyte compounds present, whatsoever
   (d) Inorganic components in the solution mix with organic components out of the solution
   (e) Grandma Moses said so!

5. Collagen fibers mainly composed of amino acids would be properly classified as:
   (a) Extracellular proteins
   (b) Intracellular carbohydrates
   (c) Membrane proteins
   (d) Phospholipid bilayers
   (e) Intracellular proteins

6. The cholesterol molecule can be considered a lipid, owing to the fact that:
   (a) It contains three fatty acids and a glycerol molecule
   (b) Cholesterol is the main blood fat that builds up on arterial walls
   (c) A high-lipid diet is, by necessity, also high in cholesterol
   (d) Cholesterol is an alcoholic steroid
   (e) Bile often contains large amounts of cholesterol

7. The glucose molecule can be considered a carbohydrate, because:
   (a) Glucose is the main large polysaccharide
   (b) Its molecular formula can be written as $(CH_2O)_6$
   (c) Fructose is a carbohydrate
   (d) Final breakdown products of protein digestion frequently include it
   (e) No other types of "sugary" molecules are organic

8. DNA and RNA differ in this way:
   (a) DNA, but not RNA, can be found within the cell's nucleus

(b) RNA has no close functional association with chromosomes

(c) RNA's five-carbon sugar has one more O atom than does DNA's sugar

(d) RNA molecules are twisted into a double helix

(e) RNA is essential for the process of protein synthesis

9. Schleiden & Schwann were more "advanced" than Robert Hooke in their thinking about the cell. Clear support for this statement comes from:

(a) Hooke's essential concepts about vitalism and homeostasis

(b) Schleiden & Schwann's published work claiming that the cell was the fundamental unit of life

(c) The historical sketches of the cork cell by Schleiden & Schwann were considerably more detailed than those drawn by Hooke

(d) Hooke's strong denial that cells were living things, at all!

(e) German universities giving more grant money to Schleiden & Schwann, compared to the mere pittance they handed out to poor Robert Hooke!

10. Glycogen is usually found at high concentrations within the liver cells of resting, well-fed athletes. This is explained by:

(a) The observation that exercise seldom requires much ATP

(b) Resting conditions don't require using much glucose for energy, while eating generally supplies more glucose

(c) Blocking of certain enzymes that should normally be operating within liver cells

(d) Overstretching of muscle tendons during periods of exercise

(e) Replacement of body fat stores with tissue proteins

11. Individual cell organelles don't have physiology, as a result of their:

(a) Lack of sufficient lipid content

(b) Small size and extreme degree of functional specialization preventing life

(c) Tendency to block mitosis, rather than promote it

(d) Ability to synthesize DNA in the absence of any messenger RNA

(e) Deficiency of structural proteins

12. The Fluid Mosaic Model was designed to describe the:

(a) Basic structure of the cell membrane

(b) Fundamental causes of skin cancer

(c) Operation of the nucleus as the cell "kernel"

(d) Physiological changes in the ribosomes over time

(e) Shifting nature of the cytoplasm in most cells

13. Both microtubules and microfilaments together make up the:
    (a) Rough ER
    (b) ECF
    (c) ICF
    (d) Cytoskeleton
    (e) Mitochondrial matrix

14. The inner mitochondrial membrane is folded into slender "crests" called:
    (a) Lysosomes
    (b) Cristae
    (c) Tendinae
    (d) Free energies
    (e) Catabolic enzymes

15. Autolysis explains why:
    (a) Human bodies don't become overstuffed with their own dead cells!
    (b) Most white blood cells don't attack one another
    (c) Foreign bacteria are often ignored by the body's immune system
    (d) Lysosomes haven't been observed, until fairly recently
    (e) Phagocytosis is a relatively rare event

16. Smooth ER's main cellular function is:
    (a) Synthesis of new cell proteins
    (b) Excretion of toxic wastes from the cell's surface
    (c) Intracellular transportation of newly constructed chemicals
    (d) Splitting of ATP
    (e) Mitosis and cleavage

17. The Golgi body primarily acts to:
    (a) Package groups of newly synthesized cell products
    (b) Phagocytose cancer or bacterial cells
    (c) Store reserve DNA outside the cell nucleus
    (d) Assist the centrioles with creating the mitotic spindle
    (e) Convert sugar into amino acids

18. The main microtubule organizing center within cells:
    (a) Mitochondrion
    (b) Cytoskeleton
    (c) Centrosome
    (d) Rough endoplasmic reticulum
    (e) Nucleolus

19. The entire life span of a particular cell:
    (a) Mitosis to cyokinesis
    (b) The Cell Cycle
    (c) The Krebs Cycle
    (d) Membrane transport
    (e) The Cell Structure–Function Sequence

20. The first phase of mitosis, when a mitotic spindle appears:
    (a) Metaphase
    (b) Anaphase
    (c) Interphase
    (d) Periphase
    (e) Prophase

21. Involves the splitting of the centromeres and separation of duplicate chromatids:
    (a) Anaphase
    (b) Postphase
    (c) Centrosplitophase
    (d) Interphase
    (e) Mayonnaise

22. Telophase can be literally translated as:
    (a) "Middle" phase
    (b) "After" phase
    (c) "End" phase
    (d) "Mitotic" phase
    (e) "Telling" phase

23. A tissue includes both a collection of similar cells, plus:
    (a) Extracellular protein matrix and plasma or interstitial fluid
    (b) A heavy concentration of hemoglobin
    (c) Invading parasites within the organism
    (d) Intracellular substance, otherwise known as cytoplasm
    (e) Abundant quantities of free or attached hairy projections

24. Areolar tissue belongs to the general or basic tissue type known as:
    (a) Epithelial tissue
    (b) Muscle tissue
    (c) Loose, "air-filled" cartilage
    (d) Tight nervous tissue
    (e) Connective tissue

25. The Principle of Cellular Pathology first stated that:
    (a) Pathological anatomy often shows up at the Chemical Level
    (b) Two cells are certainly better than one!
    (c) Cells are at the very root of living organisms
    (d) For an explanation of many diseases, look for abnormalities within cells
    (e) The organelles (especially the mitochondria) always create the worst diseases!

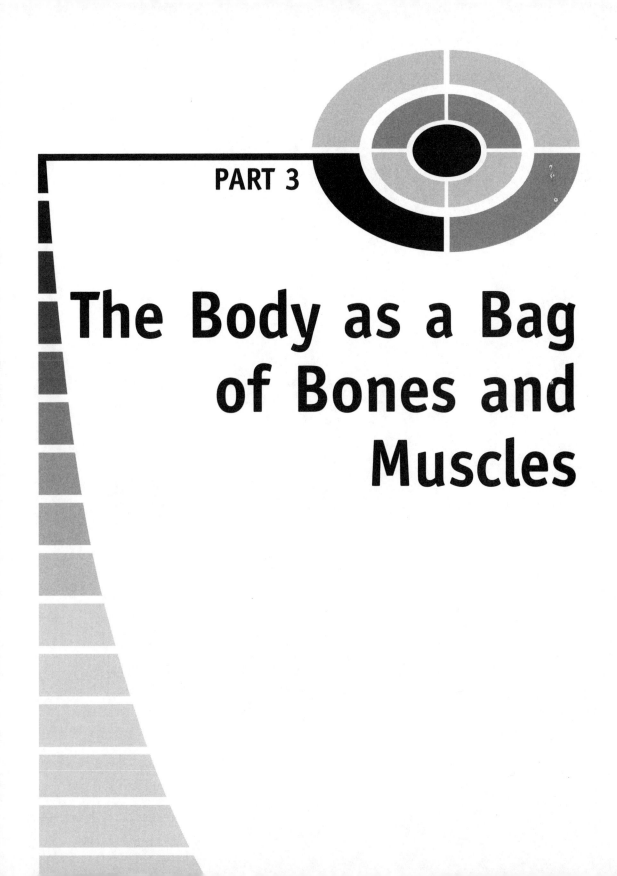

PART 3

The Body as a Bag of Bones and Muscles

6

Our Tough Hide: Anatomy of the Skin

In Part 2 of this book, the basic building blocks of all body structures were "demystified." We dug deep down, and we found a bunch of chemicals hidden in our body basement (Chapter 4). Recall that this lowest Chemical Level of body organization holds Levels I–III (Subatomic particles, Atoms, and Molecules). We then took three steps upward from the Basement (where it is always lifeless and dark), and looked at Levels IV–VI (Organelles, Cells, and Tissues). Chapter 5, you may remember, briefly defined an organ (Level VII) as:

ORGAN = A collection of two or more of the basic body tissues, which together perform some specific body function

There are, of course, many organs in the body, each with its own specific body functions. Incorporating and rising above the individual organ, we defined the Organ System (Level VIII). We now jog your memory, and reflect that an organ system is:

**ORGAN SYSTEM = A collection of related organs, which interact
together and carry out some complex body function**

Chapter 3 gave us "Flying Time" in a jet, as we quickly introduced and outlined the key organ systems filling Human Bodyspace. There are 10 major organ systems interacting together within the human organism (Level IX).

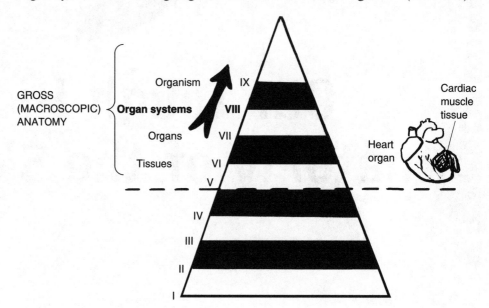

Part 3 of this book will consider three closely related organ systems – the integumentary system, skeletal system, and muscular system. "Why are we lumping these particular three organ systems together?" the curious-minded reader might ask. "Well, haven't you ever heard someone describe a really skinny, active person as just being a 'bag of bones'?" The bag is the skin or integument, the bones are in the skeletal system, and the muscular system attached to the bones, helps make this whole active "bag-of-bones" move!

The Integumentary System

The first organ system surveyed in Chapter 3 was the integumentary system. The integumentary system is our body covering, or integument (in-**TEG**-you-ment). This we defined as:

INTEGUMENTARY SYSTEM = SKIN + ACCESSORY STRUCTURES
= SKIN + HAIRS + NAILS + GLANDS

Organ System 1

For our general description in Part 3 of *ANATOMY DEMYSTIFIED*, however, we will just call the integumentary system our "Tough Hide"!

A Structural Overview of the Skin

We will use our technique of simple word equations to provide an overview of the main structural components of the skin. The names of the skin components are all centered upon a single layer – the *dermis* (**DUR**-mis). Take just a moment, and place two fingers of one hand upon the skin of your opposite hand or forearm. Now, slowly stretch the skin apart between your fingertips. Do you feel the resistance to this stretching? It is being offered by the dermis, which is the main *dense fibrous* (**FEYE**-brus) or *collagenous* (kuh-**LAJ**-ih-nus) *connective tissue* layer of the skin.

The other major skin layers are all partially named for the dermis. Specifically, these are called the *epidermis* (**EP**-ih-**der**-mis) and the *subdermis* (**SUB**-der-mis) or *hypodermis* (**HIGH**-poh-**der**-mis). Summarizing these layers, we have:

| THE SKIN | = EPIDERMIS + DERMIS + | SUBDERMIS |
| (OR INTEGUMENT) | | (HYPODERMIS) |

Organ 1

The Epidermis Says, "Pull Back My Sheets and Bedcovers!"

Take a look at someone you love. To you, they may seem beautiful. But in reality, the awful truth is this: "All that you see of someone's face is dead, horny, and waterproof!" Ridiculous, do you say? If you said so, then you have failed to understand the nature of the epidermis. The epidermis is literally the portion of the skin that lies "upon" (*epi-*) the "dermis."

HORNY, WATERPROOF BEDCOVERS: THE EPIDERMAL STRATA

The epidermis is only about as thick as this sheet of paper you are reading! Nevertheless, it consists of a series of *epidermal* (**EP**-ih-**der**-mal) *strata* (**STRAT**-uh). A *stratum* (**STRAT**-um) is a single "layer or bedcover" (*strat*) of epithelial cells. You may remember (Chapter 5) that epithelial tissue consists of tightly packed epithelial cells that either cover the body surface or line one of the body cavities.

The epidermis, then, consists of a number of stacked epidermal strata, which make the epidermis look much like a collection of thin "layers or bedcovers," heaped one upon another over a bed (see Figure 6.1). We have

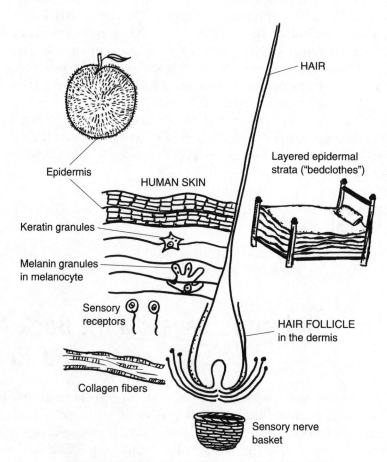

Fig. 6.1 Fundamental anatomy of the human skin.

said that the epidermis is "horny" or "hornlike" (as in the tough, waterproof "horns" of a deer or a steer). This tough, horny texture is due to the presence of a type of protein called *keratin* (**CARE**-uh-**tin**). Keratin exactly translates to mean "horn" (*kerat*) "protein" (*-in*).

Most of the epithelial cells in the upper strata of the epidermis have distinct *keratin granules* visible in their cytoplasm, when they are viewed under a light microscope. Because keratin is waterproof, and the epidermis contains so many keratin granules, the epidermis as a whole is also pretty waterproof!

The outermost layer or stratum of the epidermis consists of many flat *squamae* (**SKWAY**-me) – dead, keratin-stuffed "scales" (*squam*). Everytime you move, your clothing acts like a cheese-grater, rubbing off thousands of tiny, microscopic squamae (scales)! These fly into the air, and may get sucked up into a nearby person's nostrils! In fact, it has been estimated that up to 80% of common house dust actually consists of millions of squamae that have been shed by you, your friends, and your pet animals! A living person like you may shed about a pound of ugly squamae into the air every year! [**Study suggestion:** For confirmation of the truth of this strange fact, take a vacuum cleaner and suck up some house dust. Empty the contents of the vacuum cleaner bag, and burn it! The burning house dust (containing millions of shed, keratin-stuffed squamae) will smell much like burning chicken feathers (which are also mostly composed of keratin)!]

Tissue 1

Overall, the epidermis is classified as a *keratinized* (care-**AT**-uh-**nized**) *stratified* (**STRAT**-ih-feyed) squamous (**SKWAH**-mus) *epithelium*. This is because the epidermis is layered into epithelial strata, whose cells contain a lot of keratin. The squamous designation, which literally "pertains to scales," indicates that the topmost layer consists of flat squamae or dead "scales."

Tissue 2

"But why are the squamae *dead*?" you might ask. The reason is that, as the topmost layer of the epidermis, these cells are too far from the blood vessels in the underlying dermis to receive nutrients by diffusion. (Diffusion is a random "scattering" process, hence is only effective over very short distances.)

The epithelial cells at the base of the epidermis, however, are very much alive! They undergo frequent mitosis (cell division), and continually re-grow the epidermis from below. As the epithelial cells in the various epidermal strata are pushed higher and higher, they become progressively stuffed with more and more keratin granules. Finally, they are too far away from the dermis to receive any nutrients from the bloodstream, and they essentially become dead, keratin-stuffed scales or squamae.

MELANIN: OUR PROTECTION FROM THE SUN

Well, we now realize that the face we present to the world – our epidermis – is, indeed, dead, horny, and waterproof! But that's a good thing, isn't it? Just think of all that wear-and-tear, that rubbing and scraping, that biting cold winter wind, and that baking noonday sun! Now, speaking of the baking sun, isn't it only logical that our skin would provide us with some protection against that, too?

Our skin's protection can essentially be summed up in one word – *melanin* (**MEL**-uh-nin). Translated into English, melanin means "black" (*melan*) "substance" (*-in*). Melanin is a brownish-black pigment substance produced by *melanocytes* (**MEL**-uh-nuh-**sights**). These "black cells" are large, octopus-shaped cells with several long arms of cytoplasm (see Figure 6.1). Present near the base of the epidermis, the melanocytes appear to penetrate the membranes of adjacent epithelial cells, and inject some of their melanin granules into them. This results in a darkening of the epidermis.

Molecule 1

The chief function of melanin is absorption of *ultraviolet* (**ul**-truh-**VEYE**-uh-lit) *rays* that strike the surface of the skin. These ultraviolet (UV) rays are invisible rays whose wavelengths lie "beyond" (*ultra-*) those of X-rays, but below those of visible violet light. There is a trio of important benefits to melanin's UV-absorbing action:

1. reduction in the risk of suffering severe sunburns;
2. less risk of skin cancer (due to mutation of skin cell DNA by UV light); and
3. reduction in skin wrinkling (mainly due to the effects of UV light in creating abnormal cross-linkages of collagen fibers in the dermis).

[**Study suggestion:** In ancient times, before people migrated all over the Earth, in what part of the planet did most of the dark-skinned, dark-haired, dark-eyed populations of humans live? Where did most of the light-skinned, red-or-blonde-haired, blue-or-green-eyed human populations dwell? Can you provide a possible biological reason for this difference in population distributions?]

VITAMIN D SYNTHESIS

Surprisingly enough, even though the skin can be greatly damaged by over-exposure to ultraviolet light, some UV light is necessary for synthesis of *vitamin D*. Several of the deepest strata of epithelial cells in the epidermis react to UV light by converting a cholesterol-like steroid substance into vitamin D. Important functions of vitamin D include help in absorbing calcium and phosphorus from the intestine, into the bloodstream, and maintaining normal growth of immature bones.

EPIDERMIS SUMMARY

In summary,

> **THE EPIDERMIS = The outermost, keratinized stratified squamous epithelium covering the surface of the skin; consists of layers or strata of epithelial cells rich in keratin and melanin**

Tissue 3

The Dermis Replies, ''What – and Leave Me with a Naked Mattress?''

The dermis, as we have said earlier, is the relatively thick dense fibrous or collagenous connective tissue portion of the skin. It does, then, have many of the functional characteristics of a tough, springy bed mattress. (After all, if the epidermis is being compared to a series of bedcovers, then why shouldn't the dermis be compared to a mattress?)

The dense fibrous name comes from the dense packing of many parallel-running collagen fibers within the dermis. Collagen is known for its high degree of *tensile* (**TEN**-sil) *strength* – its ability to resist "tension" or pulling forces. The collagen fibers in the dermis (much like the fibers and springs in a real bed mattress) help maintain the shape of the skin and keep it from being overstretched.

> **THE DERMIS = The main dense fibrous connective tissue portion of the skin; has tensile strength that resists stretching of skin**

Tissue 4

The Subdermis Adds, "Don't Worry! My Fat Will Keep You Warm!"

Once the dermis and epidermis are peeled back, the subdermis (hypodermis) is exposed beneath them. The subdermis (hypodermis) is the *subcutaneous* (**sub-kyoo-TAY-nee-us**) *adipose* (**AH-dih-pohs**) *connective tissue layer*. It is subcutaneous, in that it lies "beneath" (*sub-*) the main "skin" (*cutane*) composed of the dermis plus epidermis. It is described as adipose, because it is mainly composed of *adipocytes* (**AD-ih-poh-sights**) or "fat" (*adipo*) "cells" (*cytes*).

The adipocytes store triglyceride and other lipid materials. This triglyceride store serves as a reserve food supply, in cases where the daily intake of calories is not sufficient to support life activities. It also serves as a fatty blanket of insulation, retarding heat loss from the network of blood vessels below the skin.

THE SUBDERMIS (HYPODERMIS) = Subcutaneous adipose connective tissue layer underlying the dermis; mainly consists of adipocytes

Tissue 5

Accessory Structures of the Skin

In addition to the epidermis, dermis, and subdermis (hypodermis), there are a number of accessory structures in the skin. Accessory structures are, essentially, extra things added to a particular organ system. (Think of them like the mirrors and antennas – accessories or "things added" to a car.)

HAIRS, HAIR FOLLICLES, AND SENSORY RECEPTORS

Perhaps the most noticeable of all the accessory structures in the skin is the thick forest of *hairs* that rise up from the surface of the epidermis. Each hair is basically a flexible rod of tightly packed, keratinized squamae (keratin-stuffed scales). The *hair shaft* is the portion of the hair that extends beyond the skin surface, while the *hair root* is the bottom portion embedded within a *hair follicle* (**FAHL-uh-kul**). (Review Figure 6.1.)

A hair follicle is a "little bag" lined by a membrane, and containing a hair. The base of the hair follicle lies within the dermis. But, like the epithelial cells of the epidermis, the flattened squamae of the hair are usually colored by

Tissue 6

melanin granules. There are melanocytes at the base of the hair follicle, and these normally add pigment to the squamae and color each hair as it grows.

The hairs on our bodies periodically loosen, and they are shed from their hair follicles. This periodic hair-shedding process is completely normal, and it seems to follow a definite pattern of Biological Order. Hairs in the eyebrows, for instance, are of very short duration – only 3–5 months, and then they are shed and re-grown! The hairs on our scalp, in marked contrast, often last ten times longer – 2-to-5 years – before they are shed.

Tissue 1

"I'm worried about going *bald*!" some of you might complain. "My dad was bald by the time he was 40!" "Oh, I'm sorry!" Professor Joe may respond to your understandable dismay. "I can see why you are worried about suffering from *fox mange* (**MAYNJ**)." Although it seems strange, humans who go bald are, much like foxes who lose much of their hair, suffering from *alopecia* (**al**-oh-**PEA**-she-ah), which is Greek for "fox mange"!

The problem with alopecia (baldness or "fox mange") appears to be a Breaking of the Normal Pattern of Biological Order involved in hair shedding and re-growth on the scalp. This pattern can be broken for a variety of reasons, but most important are inherited *genetic* (jeh-NET-ik) *factors* that "pertain to" (*-ic*) the "genes" (*genet*) involved in the hair growth cycle. These hair growth genes apparently turn off or inhibit the re-growth of normal-sized hairs within the follicles, after they are shed. Instead, the affected person re-grows slender "peach fuzz" hairs within the follicles. From a distance, therefore, the person whose head carries such light "peach fuzz" does, indeed, look completely bald!

Tissue 1

"What are the major functions of hairs?" the bald or non-bald reader may inquire. [**Study suggestion:** Gently run your fingertips over the hairs on one arm, without touching the skin surface. Do you feel an annoying tickling or tingling sensation?] The major function of hair is sensory reception: specifically, the feeling of touch. There is a *sensory nerve basket* around the base of each hair follicle (see Figure 6.1). When the flexible hair is bent, it agitates the nerve basket, and a sensation of touch is experienced. Overall, the dermis is absolutely loaded with *sensory receptors* of various kinds. This makes the skin the body's major organ of sensory reception. (When you kiss someone on the lips, you are pressing into their sensory receptor-rich dermis, giving you and them a thrill!)

Organ 1

NAILS AND GLANDS

There are two important groups of skin accessory structures we have not yet mentioned. These are the nails and two types of glands in the skin.

A nail as a slab of keratin

Take a glance down at the surfaces of your *digits* (**DIJ**-its). Although digits are literally "fingers" in Latin, the word also describes your toes. A nail is basically a thick slab of *stratum corneum* (**COR**-nee-um) – the hard, "horny" (*corne*) outermost "layer" of the epidermis (see Figure 6.2).

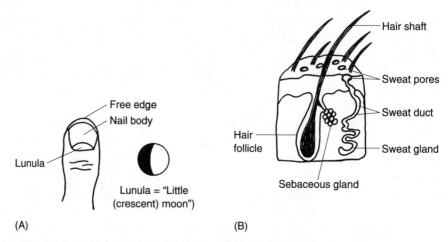

Fig. 6.2   Nails and glands: Helpful "things added" to the skin. (A) Dorsal aspect of the thumb. (B) Sweat and sebaceous glands near hair follicles.

The nail is hard and stiff because it consists essentially of thousands of dead squamae – keratin-stuffed scales – tightly pressed together into a flat bed or sheet. For all practical purposes, therefore, a nail is a tough, horn-like slab of keratin.

The *body of the nail* is the main, pinkish-colored, rectangle-shaped portion you can see on each of your digits. The pink color is created by a network of *blood capillaries* (**CAP**-ih-**lair**-eez), lying underneath. The capillaries are like pinkish "little hairs" (*capill*) filled with reddish-colored blood.

The pinkish body is bordered anteriorly and posteriorly (front-and-back) by two whitish areas. These are the *free edge* in front, and the *lunula* (**LOON**-you-lah) in the back. Both look whitish, rather than pinkish. The free edge is the curved portion extending beyond the digit. It is white due to a lack of any underlying blood capillaries. The lunula resembles a "little moon" (*lunul*) shaped like a whitish crescent. There are blood capillaries under the lunula, but the bottom layers of the epidermis are too thick in the lunula to allow us to see any pink.

"Don't sweat it! – Just get greasy!"

In addition to the nails, there are several types of glands found within the skin. Chapter 3 defined glands, and noted that there are two main varieties – exocrine glands and endocrine glands. The skin is rich in two types of exocrine glands, which are glands of external secretion of some useful product into a passageway or duct.

Sweat glands are technically called *sudoriferous* (**soo**-dor-**IF**-er-us) *glands*, because they are "sweat" (*sudor*) "carriers or bearers" (*fer*). There are about 3 million sweat (sudoriferous) glands scattered throughout the dermis (see Figure 6.2,B). Each sweat gland has a highly coiled *body*, which secretes sweat into a long *sweat duct*. Finally, the sweat duct empties into a *sweat pore* on the skin surface. Sweat is rich in water and sodium chloride (NaCl), and also contains small amounts of waste products, such as *urea* (you-**REE**-uh) and *lactic* (**LAK**-tik) *acid*. Sweat plays an essential role in body cooling and *thermoregulation* (**THER**-moh-reg-you-**LAY**-shun). By thermoregulation, we mean the "regulation" or control of body "heat" (*therm*) or temperature. The specific mechanisms of thermoregulation (control of body temperature) are discussed in some detail within our companion volume, *PHYSIOLOGY DEMYSTIFIED*.

Sebaceous (sih-**BAY**-shus) *glands* "involve or pertain to" (*-ous*) "grease" (*sebac*). The great majority of the several million sebaceous glands within the dermis are attached to the sides of hair follicles (see Figure 6.2,B). The sebaceous glands continually produce and secrete *sebum* (**SEE**-bum) or skin "grease" (*seb*). Sebum plays an often-underappreciated role in lubricating the hairs and skin surface. (Think about what happens to the skin on the hands of many people, such as nurses, who have to wash their hands often. Especially in the winter, not having enough sebum results in dry, red, painfully cracked skin!)

Quiz

Refer to the text in this chapter if necessary. A good score is at least 8 correct answers out of these 10 questions. The answers are listed in the back of this book.

1. The integumentary system consists of the:
   (a) Brain and spinal cord
   (b) Digits and toenails

(c) Skin plus accessory structures
(d) Skin plus the dermal skeleton

2. The main connective tissue portion of the skin:
(a) Epidermis
(b) Dermis
(c) Subdermis
(d) Hypodermis

3. A single "layer or bedcover" of epithelial cells:
(a) Stratum
(b) Keratinized nail
(c) The epidermis
(d) Epidermal strata

4. Granules in the cytoplasm that provide toughness and waterproofing:
(a) Melanin
(b) Mitochondria
(c) Hemoglobin
(d) Keratin

5. Humans shed about a pound of these structures each year:
(a) Hairs
(b) Sebaceous glands
(c) Squamae
(d) Follicles

6. Melanin plays a critical role in:
(a) Creating movements of the hairs
(b) Assisting with thermoregulation
(c) Coloring and protecting the epidermis from UV radiation
(d) Providing sensory information about touch to the brain

7. Chiefly consists of numerous parallel-running collagen fibers:
(a) Dermis
(b) The subcutaneous fatty layer
(c) Epidermis
(d) Stratum corneum

8. The adipocytes are mainly located in the:
(a) Hair follicles
(b) Hypodermis
(c) Keratinized stratified epithelium
(d) Sweat glands

9. A "little bag" that holds things:
   (a) Sensory receptor
   (b) Squama
   (c) Follicle
   (d) Nail bed

10. The body of a nail looks pink in color, because:
    (a) Its collagen fibers have a reddish tint
    (b) The tightly packed nail squamae are stuffed with hemoglobin
    (c) Its blood vessels are extremely fragile and bleed into the nail
    (d) Tiny capillaries are visible in the deep tissue underlying the nail

Body-Level Grids for Chapter 6

Several key body facts were tagged with numbered icons in the page margins of this chapter. Write a short summary of each of these key facts into a numbered cell or box within the appropriate *Body-Level Grid* that appears below.

Anatomy and *Biological Order* Fact Grids for Chapter 6:

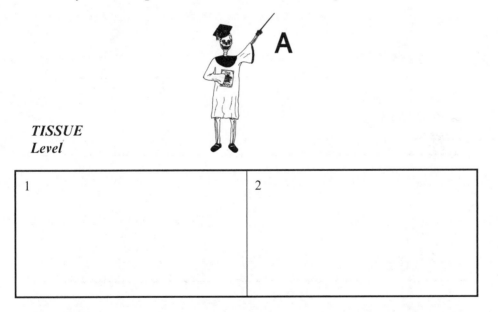

TISSUE
Level

1	2

3	4
5	6

ORGAN
Level

1

ORGAN SYSTEM
Level

1

Physiology and *Biological Order* Fact Grid for Chapter 6:

TISSUE
Level

1

ORGAN
Level

1

Function and *Biological Order* Fact Grid for Chapter 6:

MOLECULE
Level

1

Physiology and *Biological Disorder* Fact Grid For Chapter 6:

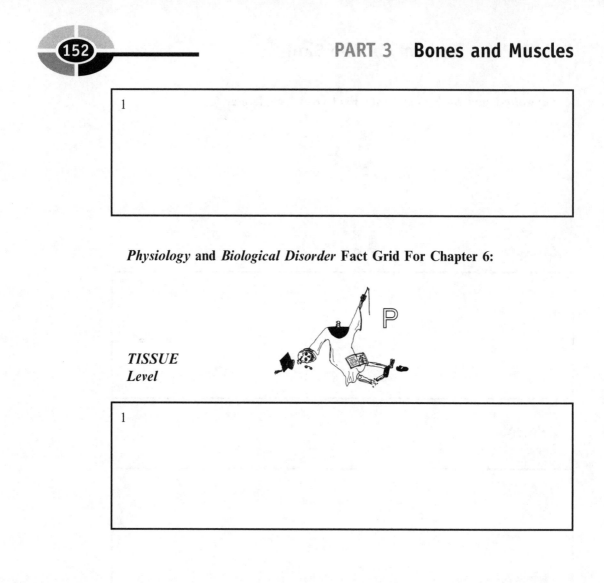

TISSUE
Level

1

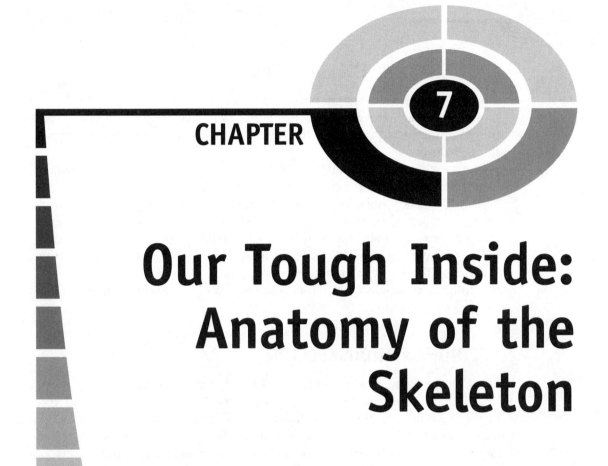

Our Tough Inside: Anatomy of the Skeleton

In Chapter 6, we talked about the skin as "Our Tough Hide." Now, in Chapter 7, we discuss the *skeleton* (**SKEL**-uh-tun) or "hard dried body." To be sure, a "hard dried body," such as a mummy, does not have much left of its body, besides its tough hide (the skin) and its tough inside (the skeleton)!

The Skeletal System

In Chapter 3, we introduced the skeleton as an organ system. Technically, it is the skeletal system, which consists of individual bone organs plus the joints made between them:

(SKELETON) SKELETAL SYSTEM = BONE ORGANS + JOINTS

THE "CRABBY" EXOSKELETON

You probably remember the phrase, *internal environment*, and the opposite phrase, *external environment*. The prefix, *endo-*, like internal, means something "inside or within." And the prefix, *exo-*, like external, denotes something "outside."

Picture a big crab, or maybe a lobster. These creatures have an extremely hard outer shell, which acts as their *exoskeleton* (**EKS**-oh-**skel**-uh-tun), or "hard dried body outside." Obviously, their exoskeleton gives them a formidable armor of protection from any predator who might wish to attack and eat their soft, delicious muscle tissue! Further, they have armored external joints between the sections of their legs. [**Study suggestion:** Go to your favorite seafood restaurant and order crab, lobster, or crab legs. Crack open the tough exoskeleton, which protects the sweet, delicate, pinkish-white skeletal muscle tissue].

THE HUMAN ENDOSKELETON

Humans are classified as *vertebrates* (**VER**-tuh-brits) – animals with spines or "backbones" (*vertebr*). And our *vertebral* (ver-**TEE**-brul) *column*, as our vertical collection of backbones, is, of course, hidden inside of our back. Since the rest of our bones are also located within the body, we have what is called an *endoskeleton* (**EN**-doh-**skel**-uh-tun) or "hard dried body within."

Let us make a useful analogy. Figure 7.1 displays the soft flesh of a peach, and its hard pit within. Similarly, the human body has its soft flesh (skeletal muscle tissue) located more externally, and the rock-hard endoskeleton lying deeper within.

The human endoskeleton consists of about 206 total bones in the adult. It has two main subdivisions. These are the *axial* (**AX**-ee-ul) *skeleton* and the *appendicular* (**ah**-pen-**DIK**-you-ler) *skeleton*. The adult axial skeleton has about 80 bones, while the appendicular skeleton contains 126 bones.

A

Organ System 1

THE HUMAN ENDOSKELETON (206 bones total)	=	THE *AXIAL* SKELETON (80 bones)	+	THE *APPENDICULAR* SKELETON (126 bones)

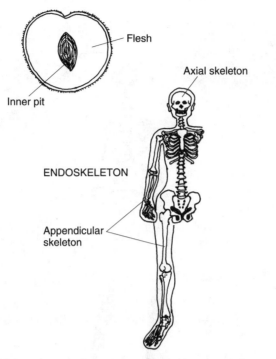

Fig. 7.1   The human endoskeleton is to the body, as a pit is to a peach!

An Overview of the Two Parts of the Skeleton

As its name suggests, the axial skeleton is the portion of the endoskeleton that lies along the body's central *longitudinal* (**long**-jih-**TWO**-duh-nal) *axis*. (Carefully study Figure 7.2.) Like the axis of the planet Earth, or the axle of two car wheels, the longitudinal axis forms the "lengthwise axle" around which the body turns or pivots. Think of the central longitudinal axis as an imaginary vertical line passing up and down through the middle of the body (much like the body midline).

The word, appendicular, literally "refers to" (*-ar*) "little attachments" (*appendicul*). The appendicular skeleton, then, is the portion of the endo-skeleton that is located within the body's *appendages* (ah-**PEN**-duh-**jes**) or "attachments." Being more precise, the appendicular skeleton consists of the bones in all four of the body appendages or limbs (two shoulders and arms, two hips and legs). Of course, the 126 bones in the adult appendicular skeleton are all directly or indirectly attached or "appended" to the bones in the axial skeleton.

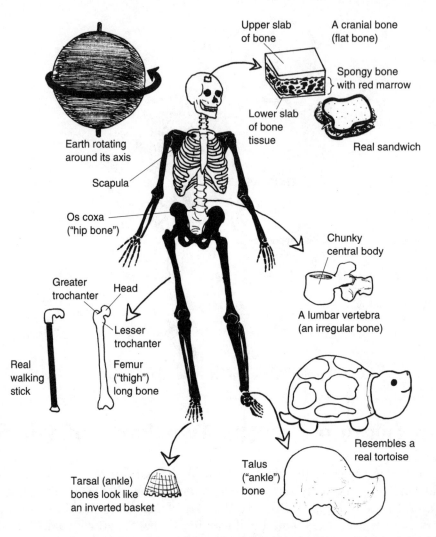

Fig. 7.2   An overview of the axial and appendicular skeletons. White = Bones of the *axial skeleton*. Black = Bones of the *appendicular skeleton*.

A

THE "WEIGHTY EIGHTY": BASIC COMPONENTS OF THE AXIAL SKELETON

Organ System 2

The axial skeleton consists of the bones of the head, neck, back, and body *thorax* (**THOH**-racks) or "chest." Specifically, these are the bones of the skull, the *middle ear bones*, the vertebral column, the *hyoid* (**HIGH**-oyd) *bone* in the front of the neck, the *sternum* (**STER**-num) "chest" or breast bone, and the *ribs*. To capsulize, we offer this handy word-equation:

$$\textbf{\textit{AXIAL SKELETON (80 bones total) or The "Weighty Eighty"}}$$

= SKULL	+ MIDDLE EAR	+ VERTEBRAL	+ HYOID	+ STERNUM	+ RIB
BONES	BONES	COLUMN	BONE		
(22)	(6)	(26)	(1)	(1)	(24)

Why do we call this group of 80 bones the "*Weighty* Eighty"? It is because many of the bones in the axial skeleton support a lot of the body's *weight*! (Isn't most of your weight found in the middle of your body?) They also play important roles in protecting the soft guts or viscera (such as the brain), and in supporting the body trunk against the force of gravity.

The shapes of some of the bones in the axial skeleton help reveal their specific functions. Each *vertebra* (**VER**-teh-**bruh**) or individual backbone, for instance, is considered an *irregular bone*. This is because it has no simple, "regular" geometric shape or form. But each *lumbar vertebra* in the "loin" area has a central *body* which is very chunky and block-like. Thus, the 5 *lumbar vertebrae* (**VER**-teh-**bray**), like a stack of solid wooden blocks, bear much of the weight of the upper body.

The hyoid is also considered irregular, but it is literally "U-resembling," and serves as a point of attachment for skeletal muscles of speech and swallowing. The *facial bones*, classified as irregular, are the more inferior skull bones that protect the delicate tissues of the *face*.

The ribs, sternum, and many of the *cranial* (**CRANE**-ee-al) *bones* in the skull, forming a roof over the brain, are classified as *flat bones*. Visualize a sandwich: two pieces of white bread, with dark meat in the middle. In a flat bone, such as a cranial bone, there is an upper and lower slab of *dense or compact bone tissue* (which are white like two slices of white bread!). "Sandwiched" in the middle, one finds a slab of *spongy or cancellous* (**CAN**-sih-**lus**) *bone tissue*, which contains small spaces filled with *red bone marrow* (i.e., looking somewhat like meat in a sandwich.) In general, flat, sandwich-like bones such as these provide a thin slab of protective armor for the soft viscera, beneath.

THE APPENDICULAR ONE HUNDRED AND TWENTY-SIX: "JUST LIKE SKINNY STICKS!"

Whereas we nicknamed the 80 bones of the axial skeleton The "Weighty Eighty," we will follow suit about nicknaming the 126 bones of the appendicular skeleton. As you can see from a quick review of Figure 7.2, most of the bones in the appendicular skeleton are classified as either *short bones* or *long bones*: some notable exceptions are the appendicular bones in the so-called *shoulder girdle*, such as the *scapula* [**SKAP**-you-lah], and the appendicular bones in the *hip girdle*, such as the *os coxa* [**AHS KAHKS**-ah], which are classified as flat bones.

Most of the 126 bones look like either short sticks (bones that are about as long as they are wide) or long sticks (bones that are much longer than they are wide). Consider, for example, the *femur* (**FEE**-mur) or "thigh" bone. It looks like a long stick, or maybe, a walking stick with a big knob on one end, called the *head of the femur*. There are also two blunt projections near the head. The larger of these is the *greater trochanter* (**TROH**-kan-ter). (Perhaps the femur should be compared to a *running* stick, rather than a *walking* stick. Greater trochanter literally means "great runner," because it serves as the point of attachment for some important leg muscles used in running.) [**Study suggestion:** Go ahead and translate the meaning of *lesser trochanter*, which is also a *bone marking* (general bone characteristic) of the femur.]

As good examples of short bones in the appendicular skeleton, take a look at the 8 *carpal* (**CAR**-pul) *bones* of the "wrist" (*carp*), and the 7 *tarsal* (**TAR**-sal) *bones* of the ankle. The word, tarsal, originally meant "flat basket" in Greek. The 7 short tarsal bones look much like an inverted (upside-down) wicker basket. In Figure 7.2, we isolate and magnify a particularly interesting short bone in the tarsal (ankle) group called the *talus* (**TAY**-lus), which literally means "ankle." Yet, the talus has a short body with a big bump on its dorsal surface (like the shell of a small turtle), a neck, and a rounded head. So, doesn't it very much resemble a tortoise?

To capsulize the appendicular skeleton, we provide this handy word-equation:

A

Organ System 3

APPENDICULAR SKELETON (126 bones) or The 126
"Just Like Skinny Sticks!"
= BONES OF UPPER APPENDAGES + BONES OF LOWER APPENDAGES
(SHOULDER GIRDLES – 4	(HIP GIRDLES – 2 bones; LEGS
bones, ARMS – 6 bones; WRISTS	– 8 bones; ANKLES – 14 bones;
– 16 bones; HANDS – 38 bones)	FEET – 38 bones)
(64 bones total in both upper limbs)	(62 bones total in both lower limbs)

[**Study suggestion:** Glance back at the summary of the appendicular skeleton. Which part (upper extremities or lower extremities) contains more bones? What general subdivisions in both the upper and lower extremities contain the same number of bones?]

The Long Bone as a Thermos Bottle

Now that we have introduced the notion of different bone shapes, and we have contrasted irregular bones, flat bones, long bones, and short bones, it is time to perform a dissection of a long bone. This is done in Figure 7.3. If we cut up our walking (or running) stick – the femur or thigh bone – we can clearly see its gross internal anatomy.

THE SHAFT AND END-CAPS: LONG BONE AS A THERMOS BOTTLE

A

Organ 1

The gross anatomy of the femur, like any long bone, is centered upon its long, main shaft, called the *diaphysis* (die-**AH**-fuh-**sis**). The diaphysis is literally a "growth" (*phys*) "through" (*dia-*) the middle of the long bone. And capping each end of the diaphysis (main bone shaft) is an *epiphysis* (eh-**PIH**-fih-**sis**) – a "growth" present "upon" (*epi-*) the shaft. In this general pattern of order, doesn't the long bone somewhat resemble a Thermos bottle, which has a main cylinder or shaft, and a plastic cap upon both ends?

Extending this analogy, suppose that a layer of paint is peeling off the surface of a Thermos bottle. Similarly, we can view the *periosteum* (**pair**-ee-**AHS**-tee-um). The periosteum is a thin membrane present "around" (*peri-*) most of the long "bone" (*oste*). [**Study suggestion:** Go to the grocery store, or your own refrigerator, and get a piece of raw chicken. Find a long bone, and run a sharp knife along the surface. Do you see a thin, milky membrane peeling off this long bone? What is its scientific name?]

Most of the long bone (like the flat bone mentioned earlier) is covered by a relatively thin shell of dense or compact bone matrix. This is the white, rock-hard, calcium-rich portion of the long bone. It lies immediately deep to the periosteum. Here and there, the periosteum and dense (compact) bone matrix are punched through with *nutrient foramina* (fuh-**RAH**-mih-**nuh**) or "feeder holes." Each nutrient *foramen* (fuh-**RAY**-mun) is just a little "hole" that carries a *nutrient blood vessel* deep into the interior of the long bone, where it delivers nutrients to the various bone tissues.

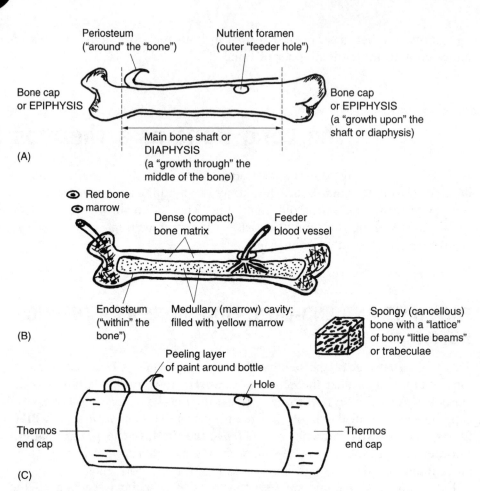

Fig. 7.3   Gross anatomy of the femur – A long bone as a Thermos bottle! (A) Gross external long bone anatomy. (B) Gross internal long bone anatomy. (C) A Thermos bottle as a familiar model.

THE SPONGY BONE AND MEDULLARY CAVITY

The interior of a long bone is displayed back in Figure 7.3 (B). Very prominent within the "middle" (*medull*) of the bone shaft (diaphysis) is the *medullary* (**MED**-you-**lair**-ee) or *marrow cavity*. The medullary (marrow) cavity is somewhat parallel to the central storage cavity found in the middle of a long Thermos bottle (Figure 7.3, C). Further, we can pretend that our Thermos bottle has its central storage cavity lined with an insulating membrane. In the case of the actual bone's medullary cavity, however, we have the *endosteum*

Organ 2

(end-**AHS**-tee-**um**). This is defined as a thin membrane "present" (-*um*) deep "within" (*endo*-) the "bone" (*oste*), as the lining of the medullary (marrow) cavity.

"Okay, so what's all this *marrow* cavity stuff, about?" the involved reader may ask at this point. The medullary cavity is alternately called the marrow cavity, because it is filled with *yellow marrow*. Named for its yellowish color, the yellow marrow chiefly consists of thousands of adipocytes. Your memory may serve to remind you (Chapter 6) that adipocytes are "fat cells" that make up adipose connective tissue. Therefore, the yellow marrow has an energy-storing physiology for the long bone.

Tissue 1

There are actually *two* types of marrow in a long bone. Remember the red marrow that was mentioned earlier, located within the spongy bone tissue of a flat bone "sandwich" in the skull (see Figure 7.2).

For long bones like the femur, most of the spongy bone tissue is located within the bone ends or epiphyses. Remember that spongy bone is also called cancellous bone. The word, cancellous, translates as "pertaining to" (-*ous*) a "lattice" (*cancell*). In spongy or cancellous bone tissue, then, there is a criss-crossing latticework of slender bone *trabeculae* (trah-**BEK**-yuh-**lie**) – "little beams" of hard bone tissue. The dark holes or gaps in between the latticework of these hard, white, beam-like trabeculae give this type of tissue the gross appearance of a sponge.

The reason that red bone marrow is colored red, of course, is that it largely consists of a soft, pulpy meshwork of red-colored blood vessels! These blood vessels of the red bone marrow snake in and out of the holes within the latticework of slender white trabeculae of spongy bone. The major function of red bone marrow is *hematopoiesis* (**he**-muh-toh-poy-**EE**-sis), which is the process of "blood" (*hemat*) "formation" (-*poiesis*). Specifically, hematopoiesis is the process of forming red blood cells, white blood cells, and blood cell fragments. The blood cells start off in the red marrow of spongy bone tissue, but they don't stay there. Instead, they circulate out of the long bone through its surface foramina (holes), within blood vessels that eventually join the general bloodstream.

Tissue 2

Bone marrow summary

Let us make an overall summary of the two types of bone marrow:

RED BONE MARROW = Mostly red-colored blood vessels within spongy (cancellous) bone tissue; functions in hematopoiesis

Tissue 1

Tissue 2

YELLOW BONE MARROW = Mostly yellow-colored adipocytes (fat cells) within the medullary (marrow) cavity of long bones; functions as an energy reserve

Bone Microanatomy

When examining a long bone like the femur with the naked eye, it appears that the spongy (cancellous) bone tissue is shot through with holes. In dramatic contrast, the compact (dense) bone tissue, as its name strongly indicates, looks solid and "hole-less," like a hard, white, solid rock!

When a thin slice of dense (compact) bone tissue is stained and viewed through a microscope, however, it looks a lot like Swiss cheese – very holey! (Examine Figure 7.4.) A series of black holes is scattered throughout a white background of *hard bone matrix*. Each black hole is called a *central* or *Haversian* (hah-**VER**-shun) *canal*. This dark canal (named after the English anatomist, Clopton *Havers*) contains a nutrient blood vessel in living dense bone matrix. Within dead, stained bone tissue, though, it is dark and empty.

Tissue 3

It is often called a *central* canal because it also looks like the black bull's-eye in the *center* of a target. There are dark pits called *lacunae* (lah-**KOO**-nee) or "lakes" that form a series of circular patterns around each central (Haversian) canal. The whole circular pattern around a central canal is called a *Haversian system or osteon* (**AHS**-tee-**ahn**). The Haversian system (osteon) is the major repeating structural subunit that makes up dense or compact bone tissue.

Characteristics Used for Naming Bones

Dense bone tissue and its Haversian systems or osteons forms a structural mainstay of practically all of the bones in the human skeleton. But things get a lot more complicated when we try to generalize about the gross anatomy of the wide variety of individual bone organs. We now describe the bone-naming characteristics used for individual bones in the skeleton, and then provide one or two specific "representative" examples:

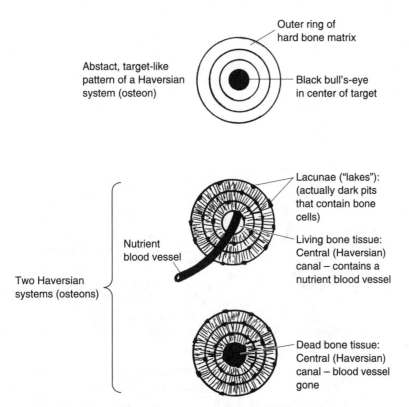

Fig. 7.4   Dense bone microanatomy: A series of Haversian systems.

Characteristic #1: Location within the body. Perhaps the most obvious characteristic used for naming bones is their location within the body. A good example of this is the *ulna* (**UL**-nah), which means "elbow" or "arm." The ulna is the bone on the little-finger side of the lower "arm," whose upper (proximal/superior) end creates the "elbow." Note in Figure 7.5, that the ulna has a very large, prominent *process* – bony projection from its surface.

This process is formally called the *olecranon* (oh-**LEK**-ruh-nahn) *process*. This name also indicates its body location as the "point or head of the elbow." [**Study suggestion:** Take advantage of the skill called *palpation* (pal-**PAY**-shun) – "the act of" (*-tion*) "touching gently" (*palpa*). Using the fingers of one hand, *palpate* (**PAL**-payt) the hard, bony, olecranon process forming the elbow on your opposite arm. This palpation procedure is a very effective method of teaching yourself the prominent parts of bone and muscle anatomy!]

Characteristic #2: Participation in creating a joint. Another characteristic that is closely related to body location is the area where a bone helps create a

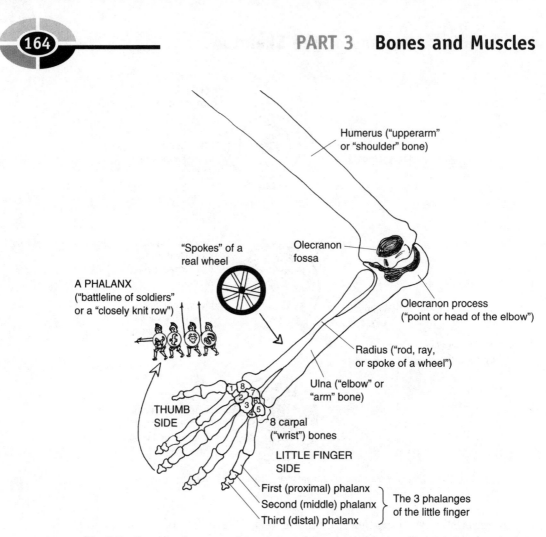

Fig. 7.5 Bones in the arm and hand named for certain characteristics (posterior view).

particular *joint*. A joint is a place of "joining" or union between two bones. A model for this way of bone naming is provided by the *humerus* (**HYOO**-mer-us). Its Latin translation means "upper arm" or "shoulder." The humerus is the long bone of the upper arm. It forms a joint at its proximal (superior) end to help create the shoulder, and a joint at its distal (inferior) end to help form the elbow joint.

Another type of bone marking involved here is the *fossa* (**FAHS**-ah) – a "ditch" or depression in the surface of a bone: think of a fossa as being the complementary opposite of a bony process. On the back (posterior) surface of the distal end of the humerus, for instance, is a deep depression called the *olecranon* (oh-**LEK**-ruh-nahn) *fossa*. [**Study suggestion:** Look again at Figure

7.5. What specific process fits into the olecranon fossa? Together, they make up what specific joint?]

Characteristic #3: Geometric shape or pattern. How about the next-door neighbor of the ulna? It is called the *radius* (**RAY**-dee-us). This bone reflects the naming of a bone after a geometric shape or pattern. Perhaps the first thing that comes to mind when we hear the word, radius, is half the distance across the diameter of a circle. The radius means "rod, ray, or spoke of a wheel" in Latin. Like the spoke of a wheel, the radius is a long, relatively thin rod of bone that lies along the thumb-side of the forearm.

Now, let us move down through the hand, and out to the fingers. Here we have one set of *phalanges* (fah-**LAN**-jeez). Each *phalanx* (**FAL**-anks) is literally a "battle line of soldiers" or a "closely knit row." As is evident from Figure 7.5, each phalanx is a small long bone in one of the fingers (or one of the toes). The pattern, here, is a very imaginative connection back to the horizontal rows and vertical columns of the phalanges (battle lines of soldiers) in Ancient Greece and Rome. Thus, each digit of the hand (except for the thumb) bears three phalanges in a fairly straight column. These are called the *first or proximal phalanx*, *second or middle phalanx*, and *third or distal phalanx*.

Bone Fractures

We have been discussing the various characteristics of normal, intact bones. Now it is time to outline the concept of *bone fracture*. A fracture is a "break" in a bone. A frequent cause of bone fracture is overstretching of bone collagen fibers, such that they snap like breaking ropes. (More pathophysiology of bone fracture is provided in *PHYSIOLOGY DEMYSTIFIED*.)

Organ 1

SIMPLE FRACTURES VERSUS COMPOUND FRACTURES

A *simple fracture* is a break in a bone, with no damage to the overlying skin. A *compound fracture* is a break in a bone that also involves damage to the overlying skin, such that the broken bone ends may or not stick out through the skin. The compound fracture obviously offers much more risk of suffering contamination of the fractured area with bacteria.

Organ 2

A COMPLETE FRACTURES VERSUS GREENSTICK FRACTURES

A *complete fracture* is a break all the way through a bone, such that two or more separate pieces result. An *incomplete fracture*, also called a *greenstick fracture*, is a break part-way through a bone. Like a green stick in the Spring, a bone with a greenstick (incomplete) fracture does not have a clean snapping break all the way through, but rather a partial break and shredding. [**Study suggestion:** Ask yourself this question: "Who would be more likely to suffer a greenstick fracture of a particular long bone – an adult, or a young child? Why?"]

COMBINATIONS OF FRACTURE TYPES

Figure 7.6 provides a summary of combinations of the above fracture types. *Simple complete fractures*, for example, involve a break all the way through a

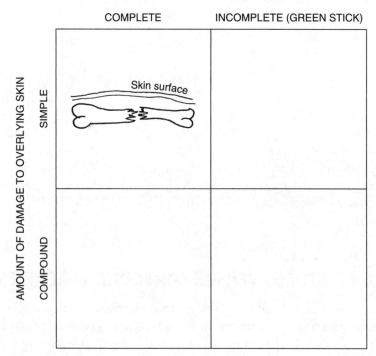

Fig. 7.6 Four general types of bone fractures. (**Note:** Only the first box or cell contains an illustration. The reader is asked to fill in the other 3 boxes in the figure.)

bone, but with no damage to the overlying skin. *Simple incomplete fractures* involve a break part way through a bone, with no damage to the overlying skin. *Compound complete fractures* entail a break all the way through a bone, accompanied by a break in the overlying skin. And *compound greenstick (incomplete) fractures* have a partial break through a bone, accompanied by damage to the overlying skin. A simple sketch of a simple complete fracture is provided in the first box in Figure 7.6. [**Study suggestion:** Now, you go ahead and sketch the other three types in their labeled boxes. Check your answers with a friend.]

Major Types of Joints

We have defined joints as places of meeting between bones. There are three main types or categories of joints within the human body (Figure 7.7).

FIBROUS JOINTS (SYNARTHROSES)

The simplest group are the *fibrous* (**FEYE**-brus) *joints* or *synarthroses* (**sin**-ar-**THROW**-seez). As their name suggests, the meeting bones in a fibrous joint are more or less strapped together by a set of collagen fibers. A fine example is provided by the *sutures* (**SOO**-churs) or jagged "seams" (*sutur*) running between individual skull and facial bones.

Organ 3

The sutures and other types of fibrous joints are immovable. This gives them the alternate name of synarthroses – literally "conditions of" (*-oses*) "joints" (*arthr*) with the bones strapped "together" (*syn-*).

CARTILAGINOUS JOINTS (AMPHIARTHROSES)

The second group of joints are called the *amphiarthroses* (**am**-fee-ar-**THROW**-seez). These are "joint conditions" (*arthroses*) permitting movement "on both sides" (*amphi-*) of the involved bones. Amphiarthroses are only partially movable joints, but (as their name states) their bones can move on both sides, and in all directions.

Organ 4

Think about the *intervertebral* (**in**-ter-ver-**TEE**-bral) *joints* that are sandwiched "between" (*-inter*) the individual vertebrae. Since the intervertebral joints are slightly movable, you are able to bend your back and twist your trunk moderately as the jointed vertebrae move short distances on both of their sides (top and bottom) and in all directions.

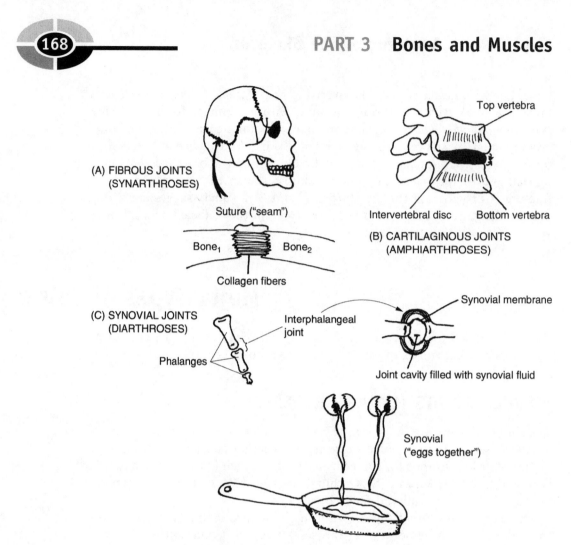

Fig. 7.7   The three main types of joints.

There is an oval slab of *cartilage* (**CAR**-tih-**laj**) *connective tissue* between the bodies of each two stacked, jointed vertebrae. This slab is termed an *intervertebral disc*. Since the intervertebral discs are composed of cartilage, the intervertebral joints including them can be classified as cartilaginous joints.

Like stale, spongy marshmallows, the collection of intervertebral discs between the vertebrae also provide excellent shock absorption!

SYNOVIAL JOINTS (DIARTHROSES)

The third group of joints are the *diarthroses* (**die**-ar-**THROW**-seez) or *synovial* (sin-**OH**-vee-al) *joints*. The word, *diarthrosis* (**die**-ar-**THROW**-sis), means

Organ 5

a "double" (*di-*) "joint" (*arthr*) "condition" (*-osis*). You probably know of certain people who are said to be "double-jointed." Such people have an unusually high degree of mobility of the *interphalangeal* (**in**-ter-fah-lan-**JEEL**) *joints* located "between" (*inter-*) their "finger or toe bones" (phalanges). Therefore, the diarthroses are the freely movable joints, with bones so movable that many do seem to be double-jointed!

"What about the synovial name?" you ask. Synovial "pertains to" (*-al*) "eggs" (*ovi*) whose whites have been poured "together" (*syn-*). This creative thinking reflects the gross appearance of the actual *synovial fluid* found within the *joint cavity*. Secreted by a *synovial membrane* lining the joint cavity, the synovial fluid is clear, thick, and slimy. Thus, it closely resembles the raw whites of many eggs that have been poured into a frying pan together! The slippery and slimy nature of synovial fluid enables it to significantly reduce bone friction and wear while the body carries out most of its major movements.

Arthritis: An Inflammation of the Joints

We are rarely conscious of our joints, and frequently take for granted their amazing ability to keep our bodies moving without noticeable pain or friction. But this is certainly not the case for people afflicted with *arthritis* (arth-**RYE**-tis)!

Arthritis is an "inflammation of" (*-itis*) the "joints" (*arthr*), usually accompanied by pain and swelling. Two representative types are *osteoarthritis* (**ahs**-tee-oh-ar-**THREYE**-tis) and *rheumatoid* (**ROO**-mah-toyd) *arthritis*.

Organ 3

Osteoarthritis, the most common form of arthritis, gets part of its name from the *degenerative* (dee-**JEN**-er-ah-tiv) changes that occur within the inflamed joints. The *articular* (ar-**TIK**-you-lar) *cartilage* at the ends of bones, helping to form the "joint" (*articul*), is frequently broken down. There are secondary, pathological changes, such as an abnormal increase in size, within the "bone" (*osteo*) tissue underlying the joint. (This explains the first part of the name, *osteo*arthritis). Osteoarthritis (abbreviated as *OA*) is common wear-and-tear arthritis. It is often associated with chronic overuse, or improper use, of particular joints. Consider, for instance, the frequent occurrence of OA within the wrist joints of professional meat-cutters.

Rheumatoid arthritis is a severe *autoimmune* (**AW**-toh-ih-**myoon**) *disorder*. The person's immune or self-defense system mistakenly manufactures *autoantibodies* (**AW**-toh-**AN**-tih-bah-deez) – abnormal proteins that are produced

"against" (*anti-*) a person's "own" (*auto-*) joint tissues. The autoantibodies chemically attack and destroy the synovial membrane of certain joints, particularly those in the fingers and toes. The result can be a severe and crippling deforming of the digits.

Quiz

Refer to the text in this chapter if necessary. A good score is at least 8 correct answers out of these 10 questions. The answers are listed in the back of this book.

1. Crabs have one, but humans don't:
   (a) Keratinized integument
   (b) Endoskeleton
   (c) Skeletal system
   (d) Exoskeleton

2. Humans are classified as vertebrates. This means that their bodies contain:
   (a) An endoskeleton
   (b) Cartilage connective tissue
   (c) A series of backbones
   (d) Epidermal strata

3. All of the bones lying along the longitudinal axis:
   (a) Cranium
   (b) Axial skeleton
   (c) Pelvis
   (d) Appendicular skeleton

4. The sternum is part of the:
   (a) Appendicular skeleton
   (b) Middle ear group
   (c) Skull bones
   (d) Axial skeleton

5. Flat bones consist of:
   (a) A sandwich of red marrow between two slabs of dense bone tissue
   (b) Groups of digits
   (c) A ball of cartilage surrounding a gooey sphere of chocolate
   (d) Flat sheets of stratified squamous epithelium

6. There are ____ bones in the appendicular skeleton, but ____ in the axial:
    (a) 159; 124
    (b) 126; 80
    (c) 95; 63
    (d) 126; 127

7. The 8 carpal bones are found in the:
    (a) Thigh
    (b) Knee
    (c) Ankle
    (d) Wrist

8. The main shaft of a long bone:
    (a) Diaphysis
    (b) Talus
    (c) Epiphysis
    (d) Medulla

9. The lining found within the medullary cavity:
    (a) Periosteum
    (b) Articular cartilage
    (c) Endosteum
    (d) Zoonosis

10. The cartilaginous joints are also called:
    (a) Synarthroses
    (b) Diarthroses
    (c) Sutures
    (d) Amphiarthroses

Body-Level Grids for Chapter 7

Several key body facts were tagged with numbered icons in the page margins of this chapter. Write a short summary of each of these key facts into a numbered cell or box within the appropriate *Body-Level Grid* that appears below.

Anatomy and *Biological Order* Fact Grids for Chapter 7:

A

TISSUE
Level

1	2

3

ORGAN
Level

1	2

3

4

5

ORGAN SYSTEM
Level

1

2

3

Physiology and *Biological Order* Fact Grids for Chapter 7:

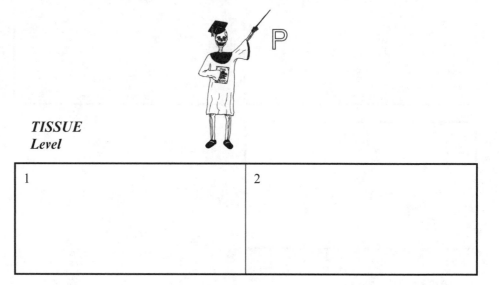

TISSUE
Level

1	2

Anatomy and *Biological Disorder* Fact Grids for Chapter 7:

ORGAN
Level

1	2

3

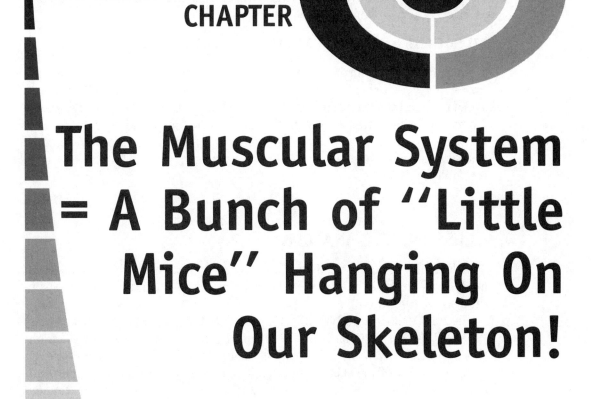

The Muscular System = A Bunch of "Little Mice" Hanging On Our Skeleton!

Part 3 of *ANATOMY DEMYSTIFIED* described the human body as "A Bag of Bones and Muscles." Well, in Chapter 6 we did the "Bag" (skin or integument), and in Chapter 7 we did the "Bones" (skeleton). Now it is time to finish up with what is inside the "Bag" and around the "Bones"! We mean, of course, the muscles!

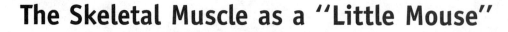
The Skeletal Muscle as a "Little Mouse"

Chapter 3, you may remember, introduced the muscular system and summarized it by the word-equation:

MUSCULAR SYSTEM = SKELETAL MUSCLE ORGANS + TENDONS

And Figure 3.1 further talked about the biceps brachii muscle in the upper arm as being representative of the approximately 600 different skeletal muscle organs. It depicted the biceps brachii according to the Common English translation of the Latin term, muscular. We know that it literally "pertains to" (-*ar*) a "little mouse" (*muscul*). The tough tendons hooking skeletal muscles onto bones, moreover, were seen to somewhat resemble the tails of mice!

GROSS MUSCLE ANATOMY

This light introduction needs to be followed up with a much more in-depth analysis of the gross anatomy of skeletal muscles. We can begin to do this in tandem with Figure 8.1.

Figure 8.1 shows us a typical *fusiform* (**FYEW**-zih-form) or "spindle" (*fusi*) "shaped" (*-form*) muscle. It has been cut in a cross (transverse section), right through the *muscle belly*. The belly is the central "bag" – the thick, bulging, middle portion of the skeletal muscle. The belly is contrasted with the *tendinous* (**TEN**-dih-**nus**) *portions* of the muscle. These are the thinner, tapered ends on either side of the belly, where the muscle eventually merges with one or more tendons.

The muscle fascia

Tissue 1

The skeletal muscle is not naked, in the sense of not having any coverings: rather, there are a number of *muscle fascia* (**FASH**-ee-uh) – "bands or sheets" (*fasci*) of fibrous connective tissue that surround or penetrate the muscle. The most prominent of these fascia is the *epimysium* (**ep**-uh-**MIS**-ee-um). As its name reveals, the epimysium is a milky covering of fascia that is "present" (*-um*) "upon" (*epi-*) the entire "muscle" (*mys*). [**Study suggestion:** Look at a raw piece of chicken. Peel the skin back slightly from the flesh, and you will

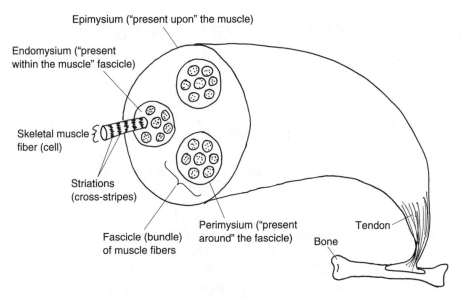

Fig. 8.1   The gross anatomy of a fusiform muscle.

see a milky-looking membrane lying upon the meat. What is the name of this membrane?]

Examining the muscle's interior, there are a several *fascicles* (**FAS**-uh-kuls) evident. A fascicle is a "little bundle" of muscle fibers that is surrounded by sheets of fascia. The *perimysium* (**pair**-uh-**MIZH**-ee-um) is the fascia present "around" (*peri-*) each bundle or fascicle of muscle fibers.

Probing still deeper, there is an *endomysium* (**en**-doh-**MIZH**-ee-um) located "within" (*endo-*) each bundle or fascicle, and between its individual muscle fibers.

You will note from Figure 8.1 that the skeletal muscle fibers, themselves, are *striated* (**STRY**-ay-tid) or "furrowed." By this, we mean that they are cross-striped with blackish lines.

Finally, a single skeletal muscle fiber is shown projected part-way out of its bundle (fascicle). A skeletal muscle fiber is actually a living, slender, "fiber"-like muscle *cell*. (In reality, you cannot see a single muscle fiber with your naked eyes, since it is microscopic.)

We will not discuss any smaller or deeper structures within the skeletal muscle fiber, in this book. For an in-depth discussion of muscle fiber micro-anatomy and physiology, please see the companion book *PHYSIOLOGY DEMYSTIFIED*.

Naming the Skeletal Muscles – A Fun Exercise in Imagineering!

In this humble volume, of course, we cannot possibly dissect all 600 individual skeletal muscles! There are just too many of these organs hanging onto our skeleton like reddish-colored mice! What we can do, however, is to provide a listing of the *characteristics* used for *naming* various skeletal muscles.

These muscle-naming general characteristics can often be applied numerous times, in combination with other characteristics, to name the specific skeletal muscle organs:

Characteristic #1: Number of muscle "heads". We have started out focusing our attention upon a particular muscle – the biceps brachii. The first part of this name, *biceps*, literally means "two" (*bi-*) "heads" (*ceps*). A head or *cep* is a major division of a muscle that has its own attached tendon. In Figure 8.2,

Organ 1

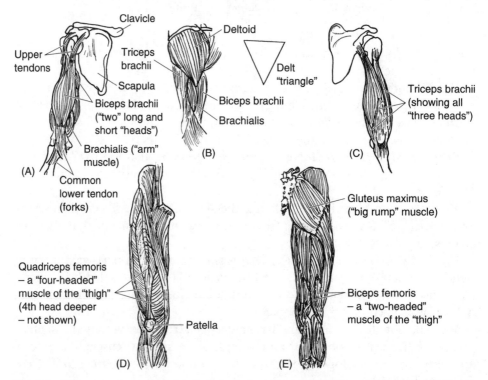

Fig. 8.2 Some muscles named for their body location and number of "heads." (A) Upper arm (anterior view). (B) Upper arm and shoulder (lateral view). (C) Upper arm (posterior view). (D) Thigh (anterior view). (E) Thigh (posterior view).

for instance, we see that the biceps brachii has two heads (ceps) – a *long head* and a *short head.* The same is true for another skeletal muscle pictured, the *biceps femoris* (**FEM**-or-is). Observe that the two heads of each of these muscles have their own tendons hooking them onto the skeleton.

There are muscles with *more than* two heads. If you study Figure 8.2 again, you see that there is a *triceps* (**TRY**-seps) *brachii* muscle, as well as a *quad-riceps* (**QWAD**-rih-seps) *femoris* (**FEM**-or-is) muscle group. [**Study sugges-tion:** Using your knowledge gained from analyzing the biceps muscles, write out or say the English translation of triceps, and then do this for quadriceps. Can you see some reinforcement for this naming in Figure 8.2?]

Characteristic #2: body location. Now that we have solved the first part of the naming mystery for biceps brachii, what about the second half, *brachii*? The word, brachii, means "arm." And the word, *femoris*, indicates "presence of" (-*is*) the "thigh or femur" (*femor*). Consequently, muscles whose last name ends in *brachii* are located in the arm, while those whose last name ends in *femoris* are located in the thigh, along the femur bone. [**Study suggestion:** Put together Characteristics #1 and 2, and you get muscles named for both their number of heads, and their body location. Which of the muscles shown in Figure 8.2, for example, is literally a "two-headed" muscle in the upper "arm"? Which is a "four-headed" muscle located in the front of the "thigh"?]

Characteristic #3: Points of attachment to the skeleton. All skeletal muscles are attached to certain points on the skeleton by their tendons. But there is usually a big difference between the degree of movement that one tendon or attachment of a muscle permits, as compared to the other tendon. This gets us into the concept of *origin* versus *insertion.*

The origin is the *least movable* tendon or attachment of a muscle. Conversely, the insertion is the *more movable* tendon or attachment. You can tell which end of a particular muscle is the origin, and which the insertion, without always having to memorize this information for every single muscle if you learn this important **Muscle Freedom-of-Movement Rule:**

Organ 1

> During contraction, the *insertion* end of a muscle *moves towards* the *origin* end.

It may be helpful to imagine being in a small rowboat. At the stern or back-end is the anchor. This anchor (for a skeletal muscle) is its origin end. Now, when you lower the anchor into the water and start fishing, doesn't the stern remain stationary (being the origin), while the bow (front end) is freely moving in the wind? The bow is thus the insertion end. Therefore, a strong wind will blow the bow (insertion) around in a half-circle, back towards the stern and its anchor. Hence the insertion end (bow) moves towards the origin end (stern with its anchor).

Let us be more anatomical and physiological, and apply this new information to our old friend, the biceps brachii muscle. The biceps brachii is an important *flexor* (**FLEKS**-or) or "bender" of the forearm. Review Figure 8.2, and observe that the superior heads of the biceps brachii are attached to bony processes along the top of the *scapula* (**SKAP**-you-lah), commonly called the "shoulder blade." Conversely, the inferior heads merge into a common tendon, which hooks onto the radius in the forearm. Now, which end is the origin of the biceps brachii, and which end is the insertion? [**Study suggestion:** Go ahead and flex one of your forearms. Remember the Muscle Freedom-of-Movement Rule. Which end does the moving? That one is the insertion!]

If you are able to observe the actions of a particular muscle (such as the biceps brachii), then you can often logically figure out which end is the insertion, and which the origin, simply by close watching. There are a few skeletal muscles, though, that are actually named for their origin and insertion.

A good example of this group is the *sternocleidomastoid* (**ster**-noh-**kleye**-doh-**MASS**-toyd) muscle. The sternocleidomastoid is a long, strap-like muscle that runs along the side of the neck (Figure 8.3). The first part of its name,

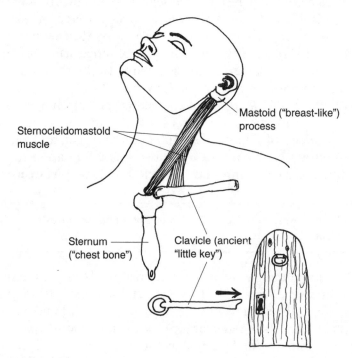

Sternocleidomastoid muscle

Mastoid ("breast-like") process

Sternum ("chest bone")

Clavicle (ancient "little key")

Fig. 8.3 The sternocleidomastoid muscle and its attachments.

stern, indicates that it hooks onto the *sternum* (**STER**-num) – the breastplate or central bone of the "chest." The second root or idea, *cleid*, represents the *clavicle* (**KLAV**-uh-kul). Known by laymen as the collar bone, the clavicle is actually named for its resemblance to a "little key" (*clavicul*) used to open heavy doors in ancient times. Finally, the superior end of the sternocleidomastoid hooks onto the *mastoid* (**MASS**-toyd) *process* of the skull. You can palpate (feel with your fingertips) the mastoid process as a hard, "breast-like" projection just posterior to your ear.

When the sternocleidomastoid contracts, it nods the head, drawing the chin down upon the chest. [**Study suggestion:** Use your newly-acquired practical knowledge to answer this question: "Which end of this muscle is the insertion? Is it the sternum and clavicle end, or is it the mastoid process end?"]

Characteristic #4: Muscle shape. Various skeletal muscles are imaginatively named for their shapes, which may resemble certain geometric forms or other objects. Consider, for instance, the *deltoid* (**DELL**-toyd) muscle – our fleshy shoulder pad (see Figure 8.2,B). The deltoid literally "resembles" (-*oid*) a "triangle" (*delt*).

Organ 2

Going up into the face (Figure 8.4), we see the *orbicularis* (or-**BIK**-you-**lair**-is) *oris* (**OR**-is) and the *orbicularis* (or-**BIK**-you-**lair**-is) *oculi* (**AHK**-you-**lie**). Each of these facial muscles is shaped like a "little orbit or little circle" (*orbicul*). In the case of the orbicularis oris, the muscle forms a little circle around the "mouth" (*or*). And in the case of the orbicularis oculi, the muscle creates a little circle around the "eye" (*ocul*). Both these muscles, then, are named for their geometric shape (little orbit) and their body locations (around mouth versus around eye).

These "little orbits" are generally referred to as *sphincter* (**SFINGK**-ter) muscles. A sphincter muscle is a circular muscle that *constricts* (kun-**STRICTS**) – "narrows" – or closes off, a body opening (such as in the mouth or eye). [**Study suggestion:** Imagine a sphincter muscle being like a circular drawstring around the opening of a cloth bag. When the drawstring (sphincter muscle) contracts, the opening of the bag is narrowed or closed off. But when the drawstring (sphincter muscle) relaxes, the opening of the bag is widened.]

Characteristic #5: Major body actions. Another feature used in naming muscles comprises their major body actions. Two very common, antagonistic body actions are *flexion* (**FLEK**-shun) versus *extension* (eks-**TEN**-shun). Whereas flexion is the "process of bending" part of the body, extension is the "process of straightening" part of the body.

A flexor muscle, therefore, is literally "one that" (-*or*) "bends" (*flex*). And an *extensor* (eks-**TEN**-sor) muscle is "one that" (-*or*) "straightens" (*extens*).

Organ 2

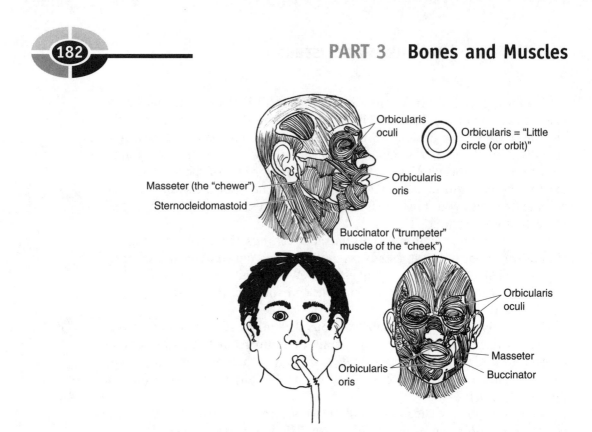

Fig. 8.4   Some representative muscles of the shoulder and face.

Such bending and straightening actions are common in the arms, wrists, hands, and legs.

A muscle named for its "bending" action is the *flexor carpi* (**CAR**-pea) *radialis* (**ray**-dee-**AL**-is). A next-door neighbor is the *extensor carpi* (**CAR**-pea) *radialis* (**ray**-dee-**AL**-is). Obviously, this muscle is partly named for its "straightening" action. (Study Figure 8.5, and note that a forearm muscle in-between these two muscles, called the *brachioradialis* [**bray**-kee-oh-**ray**-dee-**AL**-is], has been removed to provide greater clarity.)

Using our knowledge of bone anatomy (Chapter 7), do you remember the radius on the thumb side of the forearm, and the ulna on the little-finger side? (Look back at Figure 7.5, for a refresher.) And how about the carpal bones of the "wrist" (*carp*)?

Putting all of this information together, we can conclude that the *flexor carpi radialis* muscle "bends" (*flex*) the "wrist" (*carp*), and that its long lower tendon slants diagonally across the forearm to approach the "radius" bone (*radi*). [**Study suggestion:** Put down your book, and sharply flex or bend your wrist towards your body trunk. Do you see a couple of big, hard tendons popping out from your skin? Select the tendon more on the thumb side,

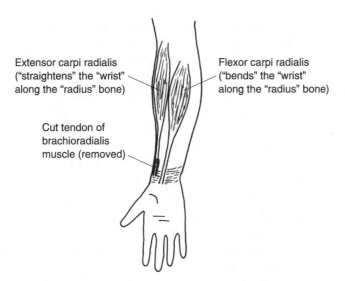

Extensor carpi radialis
("straightens" the "wrist"
along the "radius" bone)

Flexor carpi radialis
("bends" the "wrist"
along the "radius" bone)

Cut tendon of
brachioradialis
muscle (removed)

Fig. 8.5 Some flexors and extensors of the wrist (anterior view of forearm).

which is the lower tendon of the flexor carpi radialis. Now, using a finger from your other hand, palpate for the *radial pulse* in the dent made in your forearm.]

Analyzing the *extensor carpi radialis*, we note that this muscle "straightens" (*extens*) the "wrist" (*carp*), and that it runs alongside the "radius" bone.

Some skeletal muscles are named for their body actions, but more indirectly than muscles whose names include movement words like flexor and extensor. Going back to the facial muscles (Figure 8.4), we can cite the *masseter* (mas-**SEE**-ter) and the *buccinator* (**BUK**-sin-**ay**-tor).

The masseter is literally the "chewer." The masseter forms a large strap that passes down the side of the face and inserts under the bottom of the *mandible* (**MAN**-dih-bl) or "lower jaw bone." When the masseter contracts, it raises the mandible and closes the jaws. It also is important (as its name literally expresses) for chewing.

The buccinator exactly translates from the Latin for "trumpeter." The buccinator is often nicknamed the "trumpeter's muscle." It runs horizontally across the cheek, and when it contracts, it compresses (dents in) the cheek, as often occurs in trumpet players! Since *bucca* (**BUK**-ah) means "cheek," the buccinator is considered the main cheek muscle. It is also compressed when a person whistles, puckers their lips, or makes sucking motions (as in nursing infants).

Characteristic #6: Relative size. Some muscles are named for their relative size: that is, how big they are compared to their neighbors. Consider two muscles in the *gluteal* (**GLOO**-tee-al) or "rump" region – the *gluteus* (**GLOO**-tee-us) *maximus* (**MACKS**-ih-mus) and the *gluteus minimus* (**MIN**-ih-mus). [**Study suggestion:** Which do you think is the larger muscle, the one with a more "minimum," or the one with the more "maximum," size?]

Characteristic #7: Direction of muscle fibers. Several muscles are given Latin names describing the direction of their individual muscle fibers. A number of these muscles are located within the abdomen (trunk midsection). The most superficial (shallow) of these is the *external oblique* (oh-BLEEK). Figure 8.6 shows the external oblique as the "outermost" (external) muscle of the abdominal wall, with fibers that run in a "slanted" (oblique) direction. It is sometimes nicknamed the "slanted gut-pusher," because it *compresses* (pushes down upon) the viscera or "guts." Such compression may occur during lifting a heavy weight, giving birth, or straining during defecation.

You will note from Figure 8.6 that the abdominal muscles are thin and layered, like overlapping sheets of plywood. Cutting through the external oblique, the next muscle we encounter is the *internal oblique*, which has "slanted" fibers more "inside." And, slicing still further, reveals the *transverse* (tranz-**VERS**) *abdominis* (**ab-DAHM**-ih-nus). You may remember (Chapter 2) the transverse (horizontal) plane. The fibers of the transverse abdominis thus run horizontally "across" the abdominal wall.

The fourth muscle in from the skin surface is the *rectus* (**REKT**-us) *abdominis* (**ab-DAHM**-ih-nus). The rectus abdominis is a paired, strap-like muscle

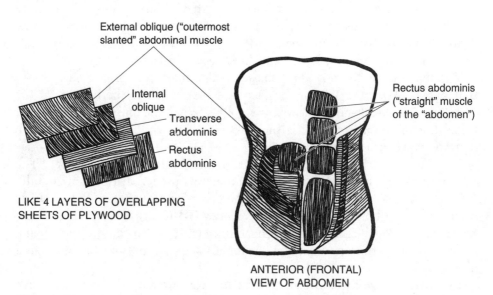

Fig. 8.6 The abdominal muscles: Four layered sheets of plywood (anterior view of abdomen).

whose fibers run "straight" (*rect*) up-and-down the middle of the "abdomen" (*abdomin*). The rectus abdominis can be nicknamed the "vertical gut-pusher" or the "sit-ups muscle." The reason is that it compresses (pushes down upon) the viscera (internal organs or "guts"). It also flexes the trunk to help a person do sit-ups.

Characteristic #8: Association with real or mythological characters. There are some muscle names associated with real people, or with people who only existed in myth or legend. A very interesting, history-related example is the *sartorius* (sar-**TOR**-ee-us). Oddly enough, this muscle exactly translates to mean "presence of" (*-us*) a "tailor" (*sartori*). But without looking at a history-related diagram (Figure 8.7), the modern reader is completely lost! The naming connection is to the way tailors used to sit in ancient times – cross-legged upon the ground while they sew. The sartorius is located along the inner aspect of each thigh. Therefore, when it contracts, it flexes (bends) the lower leg. Ancient tailors actually used the sartorius to help them sit on their bent legs!

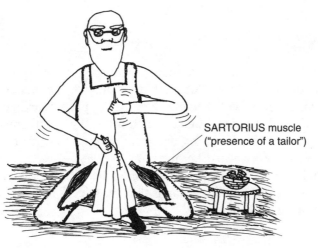

SARTORIUS muscle
("presence of a tailor")

Fig. 8.7 The sartorius: "A tailor" who sits on his legs!

Antagonistic Pairs: *Yin* and *Yang* in the World of Muscles

Speaking of Olden Times, how about Ancient China? The concepts of *Yin* and *Yang* in Traditional Chinese Medicine represent a Balance between

Opposing Forces. Just such a rough balance exists for our skeletal muscles, which are often arranged in antagonistic pairs.

A PRIME MOVER OR AGONIST

A *prime mover* or *agonist* (**AG**-on-ist) is literally a "contender," someone engaged in a contest. In Ancient Greece, agonists were the athletes who contended or competed with one another in the Olympic Games to achieve great physical goals. Somewhat similarly, prime movers or agonists are the muscles mainly responsible for contracting and carrying out a particular body movement. And whenever the agonists carry out the body movement, they are "contending" or "competing" with other muscles, that must be relaxed.

Organ 3

Returning to the arm, we can say that the biceps brachii is the main agonist or prime mover for flexion (bending) of the forearm. If this bending movement is to be successfully carried out, however, the opposing muscles that extend (straighten) the forearm must be inhibited or relaxed.

ANTAGONISTS

In short, each agonist must contend or compete with its *antagonist*. The word, antagonist, means "one that struggles against (something)." Obviously, an antagonist "struggles against" a particular agonist or prime mover. Specifically, an antagonist is a muscle that performs a movement exactly opposite to that done by a certain agonist. So, if an agonist is our *Yin*, then an antagonist must be our muscle *Yang*!

Organ 4

Going back to movements of the forearm, the triceps brachii is an antagonist of the biceps brachii. The reason is that the triceps brachii muscle extends the forearm, instead of flexing it.

SYNERGISTS

The biceps brachii does not have to struggle against or contend with the triceps brachii all by itself. The biceps brachii (unlike many other skeletal muscles) does have a helper! Its main helper is the *brachialis* (**bray**-kee-**AL**-is) muscle. [**Study suggestion:** Analyze the name, brachialis. For which of the 8 different muscle characteristics is it named?]

Go back to Figure 8.2, and examine it closely. Do you see part of the brachialis muscle peeking out at you, located deep to the biceps brachii? The brachialis is a *synergist* (**SIN**-er-jist) of the biceps brachii. A synergist is literally "one that specializes in" (*-ist*) "working" (*erg*) "with" (*syn-*) something else. In general, a synergist is a muscle that works with and assists a particular agonist or prime mover in carrying out some body movement. In the case of forearm flexion, then, the brachialis (synergist) works together with the biceps brachii (agonist) to bend the forearm at the elbow.

Organ 5

YIN–YANG PRINCIPLE OF OPPOSING MUSCLE ACTION

Most skeletal muscles in the human body are arranged into pairs of antagonists. The biceps brachii forearm flexor versus the triceps brachii forearm extender is one example of an antagonistic muscle pair. "Can the biceps brachii ever be considered the antagonist?" Yes, it can, if the body movement in question is forearm extension, rather than forearm flexion. In that case, the triceps brachii is the agonist, while the biceps brachii is the antagonist. Therefore, the labels, agonist versus antagonist, are all relative. Their use depends upon what particular body movement is being considered at the time.

We can summarize the above information by stating the ***Yin-Yang Principle of Opposing Muscle Action***:

> **Whenever a particular agonist and its synergist are excited and contract, its antagonist is inhibited and relaxed.**

Organ 6

This simultaneous excitation of agonists on the one hand, but inhibition of their antagonists on the other hand, is an impressive feat performed by the nervous system. We know that the Yin–Yang Principle is important, because whenever both an agonist and its antagonist are excited and contract at the same time, they tend to produce a body-lock or rigidity of the affected part. In the case of the forearm, it will tend to lock into mid-position, neither completely flexing, nor completely extending.

Pathological Anatomy of Muscle: The Tragedy of Muscular Dystrophy

When we stated the Yin–Yang Principle of Opposing Muscle Action, we assumed that both the agonist and antagonist muscles were normal and

Tissue 1

A healthy. Most unfortunately, such is not the case in persons afflicted with *muscular dystrophy* (**DIS**-troh-fee), abbreviated as *MD*.

Any dystrophy in the body represents a type of "bad" (*dys-*) "nourishment" (*trophy*) or a defect in metabolism. Muscular dystrophy is actually a group of inherited diseases that are characterized by severe *atrophy* (**AH**-troh-**fee**) of the skeletal muscle fibers. The word, atrophy, translates to mean "lack of" (*a-*) stimulation or "nourishment" (*trophy*). The "nourishment" in this case is not anything associated with eating a healthy diet! Rather, there are some inherited defects in the metabolism of affected muscle cells that causes them to progressively shrink and die, and eventually be replaced by fatty adipose connective tissue.

When a related group of forearm flexors (such as the biceps brachii and brachialis muscles) are hit with a severe wasting away of their fibers, then the actions of their antagonists (the forearm extenders) are not opposed. In this example, the forearm would tend to be strongly extended most of the time, and become progressively less able to bend.

Quiz

Refer to the text in this chapter if necessary. A good score is at least 8 correct answers out of these 10 questions. The answers are listed in the back of this book.

1. There are approximately ____ different skeletal muscles in the body:
   (a) 206
   (b) 350
   (c) 460
   (d) 600

2. The belly of a muscle is its:
   (a) Entire fusiform shape
   (b) Abdominal area
   (c) Tendinous portion
   (d) Thick, bulging middle portion

3. The fascia present upon an entire skeletal muscle:
   (a) Epimysium
   (b) Endomysium
   (c) Perimysium
   (d) Fascicle

4. *Pectoral* (PEK-toh-ral) literally "pertains to the chest or breast." The *pectoralis* (**pek**-tor-**AL**-is) *major* muscle, therefore, is named for what two characteristics?
   (a) Number of heads and relative size
   (b) Body location and relative size
   (c) Points of attachment and geometric shape
   (d) Resemblance to mythical characters and direction of fibers

5. The *soleus* (**SOH**-lee-**us**) muscle, located deep within the calf, is involved in *plantar* (**PLAN**-tar) *flexion* of the foot, which is bending of the sole downwards. Hence, the soleus is named for its:
   (a) Geometric shape
   (b) Resemblance to the sole of the foot
   (c) Number of heads or major divisions
   (d) Body action

6. *Trapezium* (trah-**PEA**-zee-um) means "a little table." This would most likely describe the shape of which of the following muscles?
   (a) Pisiform
   (b) Quadratus inferior
   (c) Splenius capitis
   (d) Trapezius

7. The *sacrospinalis* (**say**-kroh-spy-**NAL**-is) muscle group attaches to the *sacrum* (**SAY**-crum) at its inferior end, and along the *spine* at its upper end. When this group contracts, the spine is powerfully raised from a flexed (bent) position, into an extended (erect or upright) position. Thus, the insertion of this group is on the:
   (a) Vertebral column
   (b) Sacrum
   (c) Chest
   (d) Ankle

8. The *tibia* (**TIB**-e-ah) or "shinbone" is closely associated with the *tibialis* (**tih**-be-**AL**-is) *anterior* muscle. The tibialis anterior is named for its:
   (a) Relative size
   (b) Number of heads
   (c) Mythological connections
   (d) Body location

9. The tibialis anterior muscle causes *dorsiflexion* (**DOR**-see-**flek**-shun) of the foot, bending the top of the foot backwards. The *gastrocnemius*

(gas-trahk-**NEE**-mee-us) or calf muscle causes plantar flexion of the foot. We can conclude that these two muscles are:
(a) Antagonists
(b) Affected by MD
(c) Synergists
(d) Agonists

10. The soleus (Question #5) has what functional relationship to the gastrocnemius (Question #9)?
(a) Mutually synergistic
(b) Dually inhibitory
(c) Mutually antagonistic
(d) Cut-and-run!

Body-Level Grids for Chapter 8

Several key body facts were tagged with numbered icons in the page margins of this chapter. Write a short summary of each of these key facts into a numbered cell or box within the appropriate *Body-Level Grid* that appears below.

Anatomy and *Biological Order* Fact Grids for Chapter 8:

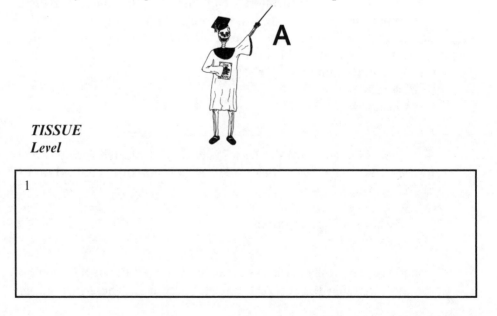

A

TISSUE
Level

1

ORGAN
Level

1	2

Physiology and *Biological Order* Fact Grids for Chapter 8:

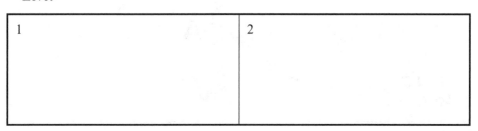

ORGAN
Level

1	2
3	4

5	6

Anatomy and *Biological Disorder* Fact Grids for Chapter 8:

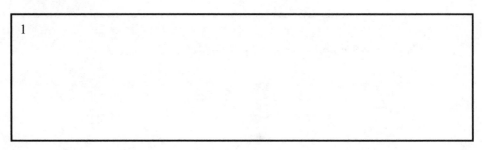

TISSUE
Level

1

Test: Part 3

DO NOT REFER TO THE TEXT WHEN TAKING THIS TEST. A good score is at least 18 (out of 25 questions) correct. Answers are in the back of the book. It's best to have a friend check your score the first time, so you won't memorize the answers if you want to take the test again.

1. Mainly consists of strata of epithelial cells:
    (a) Dermis
    (b) Hair follicle
    (c) Epidermis
    (d) Any skin accessory structure
    (e) Hypodermis

2. The horny waterproofing substance of the skin:
    (a) Melanin
    (b) Keratin
    (c) Glycogen
    (d) Calcium
    (e) Magnesium

3. When a person is exposed to ultraviolet light for a long period, the concentration of ____ in their epidermis tends to increase:
   (a) Melanin granules
   (b) Sensory nerve baskets
   (c) Keratinize squamae that are white
   (d) Glucose
   (e) Little Green Men!

4. The subcutaneous adipose connective tissue layer:
   (a) Stratum corneum
   (b) Lunula
   (c) Subdermis
   (d) Dermal collagen fibers
   (e) Endoplasmic reticulum

5. The major body function of hairs:
   (a) Sense of touch
   (b) Thickening of the skin in response to sun exposure
   (c) An attractive onset of alopecia
   (d) Provide a place for nutrients to diffuse into skin cells
   (e) When present in high numbers, serve to cool the skin surface with their shade

6. These are the real "greasers" in the skin!
   (a) Sweat pores
   (b) Hair shafts
   (c) Glycogen deposits
   (d) Pressure receptors
   (e) Sebaceous glands

7. The human endoskeleton = The ____ skeleton + The ____ skeleton:
   (a) Exoskeleton; endoskeleton
   (b) Axial; osseous
   (c) Vertebral; non-vertebral
   (d) Appendicular; somatic
   (e) Axial; appendicular

8. Flat bones are constructed somewhat like sandwiches, in that they:
   (a) Are often consumed with ketchup and mayo!
   (b) Consist of two slabs of yellow bone marrow
   (c) Usually feel like soft white bread
   (d) Have upper and lower slabs of dense bone tissue, with a slab of spongy bone between

(e) Deliver a heavy nutrient load, just like sandwiches eaten for lunch

9. The bones of the shoulder and hip girdles, arms and legs, wrists and ankles, hands and feet:
   (a) Bony vertebral column
   (b) Axial skeleton
   (c) Os coxa
   (d) Appendicular skeleton
   (e) Long bones

10. Scraping the surface of a long bone of a raw chicken with a knife, peels off the:
    (a) Marrow cavity
    (b) Epiphysis
    (c) Periosteum
    (d) Endosteum
    (e) Blastular cavity

11. A criss-crossing latticework of slender bony trabeculae:
    (a) Dense bone tissue
    (b) Articular cartilage
    (c) Haversian canals
    (d) Cancellous bone tissue
    (e) Red bone marrow

12. Functions in hematopoiesis:
    (a) Osteon
    (b) Yellow bone marrow
    (c) Spinal cord
    (d) Synovial joint
    (e) Red bone marrow

13. The major repeating structural subunit that makes up dense bone tissue:
    (a) Haversian system
    (b) Bosteon
    (c) Clopton H.
    (d) Lacuna
    (e) Central canal

14. Bone whose name means "elbow" or "arm":
    (a) Humerus
    (b) Radius
    (c) Ulna

(d) Carpal
(e) Femur

15. Line up like "battle lines of soldiers":
    (a) Phalanges
    (b) Tarsals
    (c) Vertebrae
    (d) Teeth
    (e) Carpals

16. A fracture that doesn't break all the way through a bone:
    (a) Compound
    (b) Greenstick
    (c) Complete
    (d) Simple
    (e) Complex

17. Intervertebral discs between backbones are good examples of ____ joints:
    (a) Cartilaginous
    (b) Synarthrotic
    (c) Immovable
    (d) Synovial
    (e) Diarthrotic

18. A severe autoimmune disorder attacking joint tissue:
    (a) OA
    (b) Rheumatoid arthritis
    (c) Osteomalacia
    (d) Cancer of the brain
    (e) Hematopoiesis

19. A fusiform muscle is roughly shaped like a:
    (a) Fishing bobber
    (b) Little key
    (c) Piece of gum
    (d) Round marble
    (e) Flat pancake

20. A group of muscle fibers surrounded by fascia:
    (a) Tendon
    (b) Fascicle
    (c) Epimysium
    (d) Myofibril

(e) Papilla

21. Skeletal muscle fibers are cross-striped, meaning that they are:
(a) Smooth in appearance
(b) Homogeneously dark green in color
(c) Articulated
(d) Striated
(e) Non-striated

22. Saying that, "My biceps really hurts!," is anatomically inaccurate, because:
(a) There is no such thing as a biceps!
(b) You really meant to say your triceps hurt!
(c) There are at least two different biceps muscles
(d) No muscles have both a long and short head
(e) You're just being a big crybaby!

23. "During contraction, the ____ end of a muscle moves towards the ____ end:"
(a) Origin; insertion
(b) Flagellated; non-flagellated
(c) Epimysial; perimysial
(d) Internal; external
(e) Insertion; origin

24. Forms a little circle around the mouth:
(a) Orbicularis oculi
(b) Rhomboideus major
(c) Pectoralis minor
(d) Latissimus dorsi
(e) Orbicularis oris

25. A prime mover or agonist for forearm extension:
(a) Biceps brachii
(b) Soleus
(c) Quadriceps femoris
(d) Triceps brachii
(e) Tibialis anterior

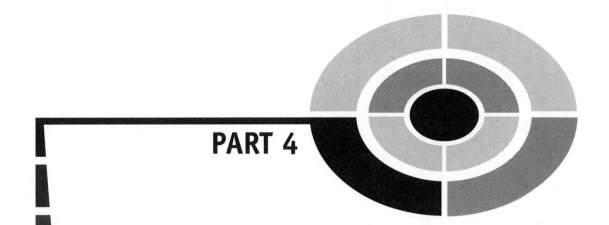

PART 4

The Nervous System and Glands: A Bunch of "Trees" and "Acorns"

CHAPTER

9

The Nervous System and Organs of the "Special Senses"

In Part 3 of *ANATOMY DEMYSTIFIED*, our focus was upon the skin, bones, joints, and skeletal muscles. After all, these organs make a major contribution to our body's existence as a Big Bag of Bones and Muscles! Collectively, we can use this information to help us define a broad concept – the *gross body soma* (**SOH**-mah). The gross body soma can be defined as the main mass of the body, which consists primarily of the skin, bones, joints, and skeletal muscles. In other words, most of the human "body" (gross or macroscopic soma) is, indeed, a Big Bag of Bones, Joints, and Muscles, covered over by the Skin! (see Figure 9.1).

In anatomy and physiology textbooks, you usually don't hear too much about the soma, until you finally get to the chapters covering the nervous system. Perhaps you have heard the phrase, "*Psyche* (**SIGH**-key) *versus Soma*." The word, psyche, comes from the Greek for "soul or mind." In

A

Organism 1

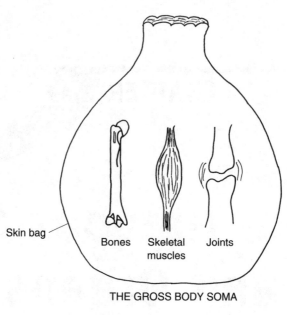

Skin bag — Bones   Skeletal   Joints
muscles

THE GROSS BODY SOMA

Fig. 9.1   A summary of the soma.

modern times, therefore, psyche is basically a word of physiology, rather than anatomy. This is because the psyche is the mind, and the mind represents the integrated functioning (i.e. physiology) of the brain.

When someone studies the anatomy and physiology of the nervous system, they will encounter the word, *somatic* (so-**MAT**-ik), quite frequently. Somatic literally "pertains to" (-*ic*) the "body" (*somat*). Specifically, somatic usually refers to structures and functions involving the gross body soma. "Where does Baby Heinie's soma come from, Professor Joe?" Ultimately, the soma comes from the embryo, as we shall soon see!

The Human Embryo Becomes Baby Heinie: Somites Create the Soma, the Neural Tube Becomes the Central Nervous System

Way back in Chapter 1, we mentioned the embryo as "a sweller." Technically speaking, the embryo represents the first 3 months of human development after fertilization. One of the early stages of the embryo is the *gastrula* (**GAS-**

true-lah). The gastrula is literally a "little stomach" (*gastrul*). It is an indented, horseshoe-shaped mass of several layers of *germ cells*. As Figure 9.2 shows, the reason that the gastrula is called a "little stomach" is that it is indented with a deep pocket (like the real stomach). This deep pocket forms the primitive digestive cavity, called the *gastrocoele* (**GAS**-truh-seal) or "stomach" (*gastr*) "cavity" (*-coele*).

THE THREE PRIMARY GERM LAYERS

There are three *primary germ layers* – three distinct layers of germ cells – within the wall of the gastrula. These cells eventually divide and *differentiate*

A

Tissue 1

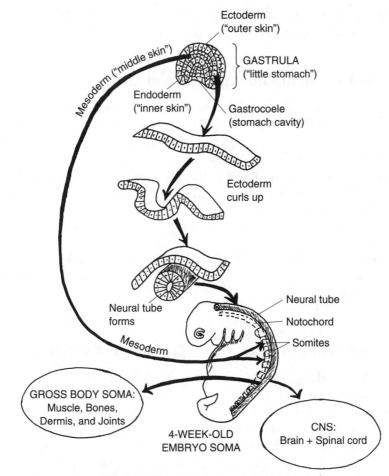

Fig. 9.2   The gastrula forms the body.

(**dih**-fer-**EN**-she-ate) or "become different" and specialize to become particular types of body tissues. (They're called *germ* cells because they "sprout" or *germinate* new cells, much like plants sprouting new growth.)

The *endoderm* (**EN**-doh-**derm**) is the "inner" (*endo-*) "skin" (*derm*) of germ cells. It lines the gastrocoele (stomach cavity) of the gastrula. Eventually, the germ cells in the endoderm differentiate to become the "inner" (*endo-*) epithelial tissue lining most of the body cavities and passageways. The endoderm also creates two important organs: the *liver* and *pancreas*. Thus, the endoderm gives rise to many of the viscera (guts).

On the outside we find the *ectoderm* (**EK**-toh-**derm**) or "outer skin." The germ cells in the ectoderm eventually create the epidermis of the skin (which, of course, is on the very "outside" of the body), as well as the accessory structures of the skin (hair, nails, and skin glands). The ectoderm also eventually forms the nervous system, including its long, branching *nerves*, as well as closely related *organs of the "special senses"* (eye, ear, etc.).

Formation of the neural tube and CNS

Figure 9.2 reveals that sheets of the ectodermal cells curl up and fuse together to form the *neural* (**NUR**-al) *tube*. The neural tube is a hollow, longitudinal (lengthwise) tube that runs along the body midline of the early embryo. The neural tube is the structure from which the entire *Central Nervous System* (*CNS*) ultimately develops. The *Central* Nervous System is formally defined as the portion of the nervous system that is *centrally* located, along the body midline. When the CNS matures, it consists of the *brain* and *spinal cord*.

To summarize the ectoderm's role in CNS development, we have:

ECTODERM $\longrightarrow$ **NEURAL TUBE** $\longrightarrow$ **CENTRAL NERVOUS SYSTEM**
(**"Outer skin"**   (**CNS**) = **Brain** + **Spinal cord**
of early embryo)

The mesoderm makes the notochord and somites

The *mesoderm* (**ME**-soh-**derm**) is literally the "middle" (*meso-*) "skin" (*derm*) of the early embryo. By the 3rd week after fertilization, the gastrula has differentiated and expanded into a recognizable soma or body. (But it looks a lot like a tadpole, at this early stage!) Much of this soma comes from the mesoderm. The developing Baby Heinie has a head, belly-stalk, and even a cute little tail! The curved soma is now stiffened by a *notochord* (**NO**-tuh-kord). The notochord develops from the mesoderm and appears as a thin "cord" (*chord*) or rod running up-and-down the little embryo's "back" (*noto-*). The notochord lies immediately deep to the neural tube (developing CNS). Eventually, the notochord matures into a full vertebral column (jointed backbone).

Budding up along the sides of the neural tube, one sees a series of *somites* (**SOH**-mights). Like the notochord, the somites develop from mesoderm. A somite is a small, cube-shaped segment or embryo "body" (*som*) "part" (*-ite*). In the 3-week-old human embryo, there is a series of these stubby body parts or somites stacked up like blocks, all along the back.

Ultimately, practically all of the gross body soma comes from the somites! One part of each somite develops into the dermis of the skin. Another part gives rise to the skeletal, smooth, and cardiac muscles. And a third part of each somite creates the bones and joints of the skeleton. So, remember this little ditty: **"FROM THE *SOMITES*, COMES THE *SOMA*!"** (The main exception, here, however, is the epidermis of the skin, which comes from the ectoderm.)

It is quite appropriate to discuss the somites in relationship to our nervous system because each somite in the embryo eventually gives rise to a tiny muscle mass, which is supplied by a *spinal nerve* (or other type of nerve).

The Peripheral Nervous System: "It's Setting My Nerves on *Edge*!"

Each spinal nerve (as well as all other types of nerves) is considered part of the *Peripheral* (per-**IF**-er-al) *Nervous System* or *PNS*. The Peripheral Nervous System (PNS) consists of nerves and *sensory receptors* that lie outside the CNS, within the body "edge" or *periphery* (per-**IF**-er-ee) (see Figure 9.3).

SENSORY NERVE FIBERS BRING INFORMATION *TOWARDS* THE CNS

Sensory receptors are modified nerve endings that are sensitive to particular *stimuli* (**STIM**-you-**lie**) or "goads" in the body's internal or external environment. *Free nerve endings* within the dermis of the skin, for example, are sensory receptors that are especially sensitive to pain and temperature stimuli. These stimuli travel as *nerve impulses* or *action potentials* (traveling waves). The action potential waves travel over *sensory* or *afferent* (**AF**-fer-ent)

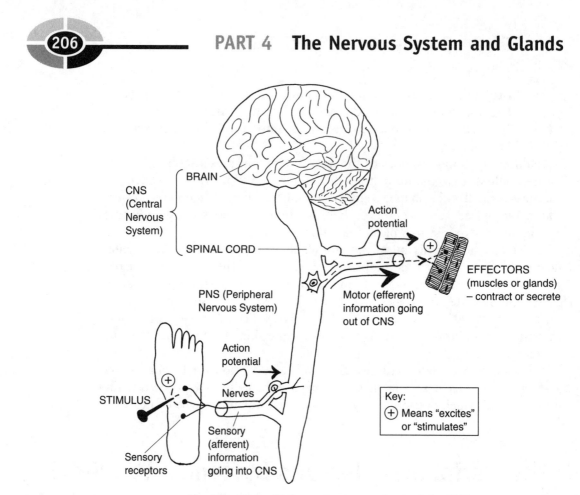

Fig. 9.3 Flow of information into and out of the CNS.

nerve fibers, "towards" (*af-*) the CNS. They provide the CNS with *sensory* information. Thus:

SENSORY RECEPTORS	$\rightarrow$	SENSORY	$\rightarrow$	CNS
(Excited by a stimulus)		(AFFERENT) NERVE FIBERS		(Receives sensory information)

MOTOR NERVE FIBERS TAKE INFORMATION *AWAY FROM* THE CNS

There is an *opposite* direction of information flow (action potential travel), too. There are thousands of *motor* ("movement"-causing) *neurons* or *nerve cells* present within the CNS. When some of these motor neurons "decide" to fire an action potential, these waves eventually travel over *motor or efferent* (**EE**-fer-**ent**) *nerve fibers*. They are called efferent nerve fibers, because they "carry" (*fer*) motor (movement-related) information "away from" (*ef-*) the CNS.

"Where do these motor or efferent nerve fibers eventually wind up, Professor?" They go out to particular *effectors* (e-**FEK**-ters). An effector is literally "something that" (*-or*) carries out a particular response, thereby having some body "effect." Specifically, the effectors in the human body are one of two types: they are either muscles that contract (causing the effect of body movement) or *glands* that *secrete* (releasing some useful substance).

CNS RELATIONSHIP TO THE PNS

"Are both these sensory or afferent nerve fibers, and these motor or efferent nerve fibers, related to specific areas of the CNS?" Yes, they may bring sensory information into, or motor information out of, either the brain or the spinal cord. We can define the brain as the *encephalon* (en-**SEF**-ah-**lahn**). The brain (encephalon) is the superior portion of the CNS. It is the part located "within" (*en-*) the "head" (*cephalon*) and skull. Conversely, the spinal cord is the narrow, "cord"-like, inferior portion of the CNS, that passes through the bony "spine."

SUMMARY OF THE CNS–PNS RELATIONSHIP

Organ System 1

Overall, we can define the *Nervous System* as the body's major system for communication and control of the internal environment (Human Bodyspace). Using a simple word equation, we have:

THE NERVOUS = CENTRAL NERVOUS + PERIPHERAL NERVOUS
    SYSTEM   SYSTEM (CNS)   SYSTEM (PNS)
            = (Brain + Spinal cord) + (Nerves + Sensory receptors)

Major Subdivisions of the PNS

As you may already be realizing, there is a huge amount of complexity built into the Nervous System! Yet, things are still highly orderly and patterned, so, if we are patient and careful, we can pretty much sort things out. (It's a lot like keeping our cool when we are trying to put together a complicated jigsaw puzzle!)

SOMATIC NERVOUS SYSTEM: SON OF THE SOMA, GRANDSON OF THE SOMITES

We began this chapter by considering the early stages of differentiation of the gastrula and its three primary germ layers into various types of body tissues. Remember that the gross body soma is essentially our Big Bag of skin, bones, joints, and skeletal muscles (as shown back in Figure 9.1). And please recall that the soma Big Bag ultimately derives from the mesoderm.

Going into more detail, we stated that the mesoderm differentiated to create the somites. The next step was that the somites, in turn, further differentiated or specialized to create the gross body soma. Now, enter the *Somatic Nervous System* or *SNS*. The Somatic Nervous System (SNS) is the portion of the Peripheral Nervous System that supplies the gross body soma. We mean that the Somatic Nervous System (SNS) picks up sensory (afferent) information from receptors in the gross body soma, and that it carries motor (efferent) information towards the effectors associated with the soma. If you feel a painful cramp in your deltoid muscle, and then rub and massage it, for instance, you are using your SNS.

Speaking a little tongue-in-cheek, we can say that the Somatic Nervous System is the *Son of the Soma* (since its nerve fibers supply the soma), and the *Grandson of the Somites* (since the soma ultimately came from the embryo's somites). Let's go right ahead and state *The Soma Rule:*

The *Somatic* Nervous System (SNS) supplies the *Soma*, which developed from the *somites*.

The Somatic Nervous System can simply be seen as the nerves going into and out of the Big Body Bag (Figure 9.4,A).

It's also the part of the PNS that is usually considered to be *voluntary* (**VAHL**-un-**tair**-ee). What we mean by voluntary, is that the movements of the bones, joints, and skeletal muscles of the gross body soma are generally under our conscious control.

THE "SPECIAL SENSES" SERVE THE BIG BODY BAG

When we are voluntarily moving our Big Body Bag (Gross Soma) down the street, how do we make ourselves *aware* of things out there that could help or hurt us? This is an important in*sight*: it is also important in smell, in taste, and in hearing! The *Organs of the "Special Senses"* – eyes (vision), ears

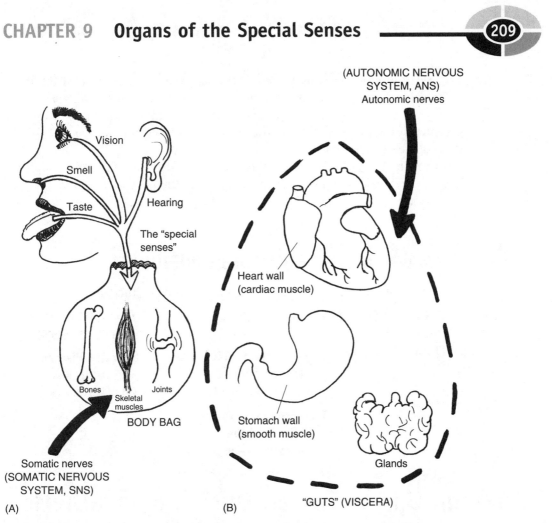

Fig. 9.4 SNS versus ANS: INSIDE the Body Bag, or OUTSIDE of it?

(hearing), tongue (taste), and nose (smell) – are also considered part of the Somatic Nervous System. These special sensory organs provide the Gross Body Soma with vital orienting information that helps it survive the rigors of life in the Real World!

AUTONOMIC NERVOUS SYSTEM: ''NO ONE TELLS *ME* WHAT TO DO!''

There is a second major subdivision of the PNS we still haven't considered. It's called the *Autonomic* (**aw**-toh-**NAHM**-ik) *Nervous System* or *ANS*. The word, autonomic, means "self-regulating."

The Autonomic Nervous System is the self-regulating portion of the PNS, in the sense that it supplies *visceral target organs* or *"gut" effectors*: i.e. it supplies smooth muscle, cardiac (heart) muscle, and glands. The Autonomic Nervous System can be seen as the nerves supplying the guts *outside* of the Big Bag (see Figure 9.4,B).

And if the Somatic Nervous System supplies the voluntary effectors, then doesn't the Autonomic Nervous System supply the *involuntary* effectors – the ones that are "self-regulating" in our guts – and beyond our conscious control? Yes.

SUMMARY OF MAJOR PNS SUBDIVISIONS

Let us now summarize the two major subdivisions of the PNS:

A

Organ System 2

PERIPHERAL = SOMATIC NERVOUS + AUTONOMIC NERVOUS
NERVOUS   SYSTEM (SNS)   SYSTEM (SNS)
SYSTEM (PNS)

Supplies the Gross   Supplies Visceral
Body Soma   Target Organs or
(Voluntary Control)   Guts (Involuntary)

Includes input from
Organs of the
"Special Senses"

"Stressed Up, or Chilled Out?": Sympathetic versus Parasympathetic Divisions of the ANS

We have already indicated that the Autonomic Nervous System supplies the guts or viscera. What we haven't done, yet, though, is say *what* the nerves of the ANS *do*, once they get there!

Since the visceral effectors or guts supplied by the ANS "do their own thing," isn't it only natural to assume that what these internal organs are doing at any given time is greatly influenced by the body's overall state of stress or arousal?

DUAL INNERVATION OF THE VISCERAL EFFECTORS

There is a *Dual Innervation* (**in**-er-**VAY**-shun) or "condition of double nerve supply" of most of the body viscera. We might compare this situation to the

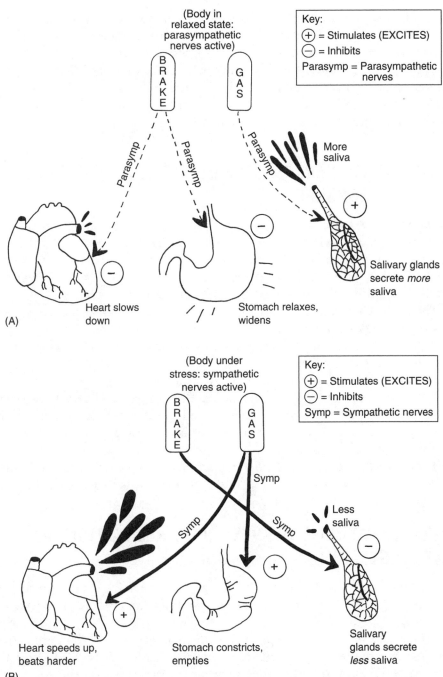

Fig. 9.5   Dual Innervation: Like the gas and brake pedals. (A) Body in a relaxed state (parasympathetic nerves active). (B) Body under stress (sympathetic nerves active).

brake and gas pedals of a car (Figure 9.5, A). The gas pedal speeds the car up, while the brake slows the car down. Both the gas and brake pedals supply the same car. They just have *opposite* effects upon the car!

A somewhat parallel, Yin–Yang influence is exerted upon particular viscera (such as the heart, stomach, and the *salivary* [**SAL**-ih-**vair**-ee] *glands*) by the *sympathetic* (**SIM**-pah-**thet**-ik) and *parasympathetic* (**PAIR**-uh-**SIM**-pah-**thet**-ik) portions of the ANS (See Figure 9.5, B).

SYMPATHETIC NERVES: STRESS AND THE "FIGHT-OR-FLIGHT" RESPONSE

You know how it feels when you are under a lot of psychological stress or pressure. You can feel your heart (cardiac muscle tissue) beating harder and faster within your chest. Your stomach wall (smooth muscle tissue) is contracting tightly into a knot. And there is very little *saliva* (sah-**LIE**-vah) or "spit" (from the salivary glands), so your mouth feels dry like cotton!

These physiological reactions form part of what is called the *Emergency Stress or "Fight-or-Flight" Response*. In this response, the body prepares itself to either stand and fight some aggressor or danger, or engage in flight and simply run away from it. In either case, the *Sympathetic Nerves* of the Autonomic Nervous System are actively engaged.

For both the heart and stomach wall, the sympathetic nerves are a gas pedal, because they stimulate (+) harder and faster contractions. But for the salivary glands, the sympathetic nerves act like a brake pedal, because they inhibit (−) the secretion of spit.

These are considered *sympathetic* reactions, because you are "suffering" (*path*) "with" (*sym*) a lot of stress, and sometimes it is really *pathetic*!

PARASYMPATHETIC NERVES: REST, RELAXATION, AND EASY DIGESTION

Now, how about your feelings when you are finally having some R (rest) & R (relaxation)? When you are very relaxed, and not under stress, it is easy for you to eat and digest food.

In this case of R & R, the *parasympathetic nerves* of the ANS are active. The parasympathetic nerves literally run right along "beside" (*para*-) the "sympathetic" nerves, and supply many of the same visceral effectors.

For example, the parasympathetic nerves act as a brake pedal for both the heart and stomach wall, because they inhibit (−) their activity. As a result, the resting heart rate is generally much lower than the active or stressed-out heart rate, and the resting stomach wall relaxes, easily letting food enter. Conversely, the parasympathetic nerves act like a gas pedal for the salivary glands, since they stimulate (+) them to increase their secretion of saliva (thereby assisting food digestion).

THE ENTERIC NERVES: GAS-VERSUS-BRAKE FOR THE INTESTINES

When you are under severe distress, there is often a very unpleasant *enteral* (**EN**-ter-al) or "intestinal" response! We are speaking about diarrhea, of course! This usually reflects the intense activity of sympathetic nerve fibers supplying the smooth muscle in the wall of the intestine.

When you feel relaxed, the parasympathetic nerves inhibit (−) the bowel wall, letting the intestines relax and fill with digested food or stool.

In recent years, anatomists have identified these mixed sympathetic plus parasympathetic nerves to the intestines as the *enteric* (en-**TER**-ik) or "small intestine" *portion* of the ANS.

Summary of the ANS

Let us now provide a capsule summary of the main portions of the Autonomic Nervous System:

AUTONOMIC = SYMPATHETIC + PARASYMPATHETIC + ENTERIC
  NERVOUS   NERVES (Active   NERVES (Active   NERVES
  SYSTEM   during stress)   during rest)   (Mixed supply
  (ANS)   of symp. &
     parasymp. nerves
     to intestine)

A

Organ System 3

The Cranial Nerves and Organs of "Special Senses"

We have now accomplished an overview of the essential portions of both the CNS (Central Nervous System) and the PNS (Peripheral Nervous System). A good place for us to start digging into some specific nervous structures is with a discussion of the 12 pairs of *cranial nerves*.

The *cranial* nerves are nerves that enter and leave the brain through holes in the skull or *cranium*. Since the cranium is the home for the brain or encephalon, which is literally "within the head (and skull)," we must look at the base of the brain to see some of these nerves. (**NOTE:** The detailed structure & function of neurons and nerve fibers is covered in the pages of *PHYSIOLOGY DEMYSTIFIED*.)

Some of the cranial nerves (numbered I through XII) are closely tied to the "Special Senses." These cranial nerves will be our main focus of attention:

Cranial Nerve Pair I: The olfactory nerve really "smells"! One of our "Special Senses" is technically called *olfaction* (ohl-**FAK**-shun) – the "process of" (-*tion*) "smelling" (*olfact*). Everything starts with the *olfactory epithelium*, which lines the roof of the *nasal* (**NAY**-sal) or "nose" *cavity* (see Figure 9.6, A).

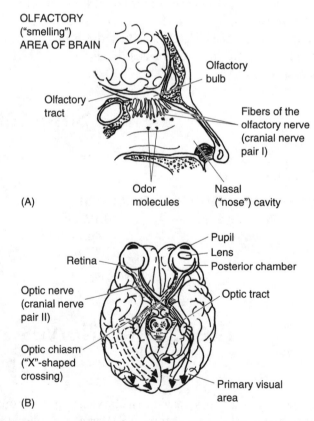

Fig. 9.6　Smelling and seeing: Cranial Nerves I & II. (A) The olfactory ("smelling") nerves (I) and associated anatomy. (B) The optic ("seeing") nerves (II) and associated anatomy.

The *olfactory nerve* is also designated as *Cranial Nerve Pair I*. The olfactory nerve has special sensory receptors, called *chemoreceptors* (**KEE**-moh-re-**sep**-tors) or "chemical receptors." These chemoreceptors are stimulated by *odor molecules* (which are, of course, chemicals) within the nasal cavity. The odor molecules are given off by anything that we can smell. The olfactory nerves carry their action potentials (traveling waves) into the rounded *olfactory bulbs*. From here, the waves go through the olfactory bulbs and enter the *olfactory tracts*. These tracts carry the olfactory information back into the brain where it is processed and recognized as certain smells.

Cranial Nerve Pair II: The optic nerve really lets us "see"! *Cranial Nerve Pair II* is named the *optic* (**AHP**-tik) *nerve*. These two nerves "pertain to vision" (*opt*). A glance at the anatomy of the eyeball is required (see Figure 9.6, B).

Light rays enter the eye through its *pupil*, the dark opening within the *iris* (**EYE**-ris), a ring that can have colors like a "rainbow." The rays are bent by the *lens* (somewhat like a camera lens). The rays continue through the *posterior chamber* of the eyeball (just behind the lens). They come to focus on the *retina* (**RET**-ih-**nah**). The retina is a "net" (*retin*) of sensory cells at the back of the eyeball. An inverted (upside-down) image is formed on the retina! (Imagine trying to read this book with all the pages inverted!)

The inverted image is changed into action potentials by the *visual receptors* in the retina. These are the *rods* (light-dark visual receptors) and the *cones* (color-sensing visual receptors). The traveling waves (action potentials) from rods and cones leave the retina through the optic nerve, which hooks into the back of the eyeball. The optic nerves from each eyeball make an "X"-shaped crossing called the *optic chiasm* (**KEYE**-asm), just below the brain.

Continuing behind the optic chiasm are the two *optic tracts*. These optic tracts penetrate the *cerebrum* (seh-**REE**-brum) or "main brain mass." The optic tracts travel all the way to the back of the cerebrum. Their destination is called the *primary visual area*. But these visual impulses are still upside-down! They just don't make any sense to us, until they are sent forward a small distance to the *visual association area*. Here the separate bits of visual information are ordered and "associated" – linked and integrated into meaningful patterns and images that we can recognize, such as the familiar face of a good friend.

Cranial Nerve Pairs VII and IX: The servers of "taste" supply the tongue. The Special Sense of taste is provided by two separate pairs of cranial nerves, but both of them supply the tongue. *Cranial Nerve Pair VII*, the *facial nerve*, runs along the side of the face (as its name suggests). The facial nerve is technically classified as a *mixed* nerve, because it carries a "mix" of two different types of information – motor information as well as sensory information. The *motor nerve fibers* stimulate the facial muscles to contract, thereby creating our various facial expressions. The *sensory nerve fibers*, however, carry taste information from the *taste buds* on the anterior (front) 2/3 of the tongue. (This is where most of our taste of sweetness is sensed.)

Cranial Nerve Pair IX is called the *glossopharyngeal* (**glahs**-oh-fah-**RIN**-jee-al) *nerve*. Its name suggests that it supplies both the "tongue" (*gloss*) and the "throat" (*pharynge*). Like the facial nerve, the glossopharyngeal is mixed. The motor nerve fibers excite the muscles of the *pharynx* (**FAR**-inks) – "throat" – to help us swallow. The sensory nerve fibrers carry taste information from taste buds on the posterior (back) 1/3 of the tongue. (This is where most of our taste of sourness and bitterness is sensed.)

In summary of the sense of taste:

TASTE: Facial Nerve (VII)   +   Glossopharyngeal Nerve (IX)
(Sweet taste, front 2/3 of tongue)   (Sour & bitter taste, back 1/3 of tongue)

Organ 1

Cranial Nerve Pair VIII: The auditory nerve gives us a fair "hearing." When you go sit in an auditorium, you expect to "hear" (*audit*) a lecture. Hence, when you see the phrase, *auditory* (**AW**-dih-**tor**-ee) *nerve* (*Cranial Nerve Pair VIII*), you know that it must "pertain to" (*-ory*) "hearing."

Figure 9.7 tells us a little about the anatomy and physiology of hearing. Sound waves enter the *auricle* (**AW**-rih-kl), the "little ear" on the outside of the head. The auricle collects the sound waves and directs them through the *Eustachian* (**you-STAY**-shun) *tube*, which is alternately called the *external auditory canal*. After they reach the end of this canal, the sound waves vibrate the *tympanum* (tim-**PAN**-um) or "eardrum."

Lying immediately deep to the tympanum (eardrum) are the three *auditory ossicles* (**AHS**-ih-kls) – the "tiny bones" of the *middle ear cavity*. As the tympanum vibrates, the first ossicle, called the *malleus* (**MAL**-ee-**us**) or "mallet" (due to its shape), is set into motion. The malleus pushes upon the flat-topped, "anvil"-like *incus* (**ING**-kus), which in turn pushes upon the *stapes* (**STAY**-peez) or "stirrup."

The stapes then pumps against the *oval window* in the *cochlea* (**KAHK**-lee-ah) or "snail shell" of the *inner ear*. The fluid within the cochlea (inner-ear snail shell) is compressed by the plunging action of the stapes. This activates the *hair cells*, which are the sensory receptors for hearing.

The hair cells are flattened by the fluid pressure, and send out a series of action potential waves onto the auditory nerve. Eventually, the auditory nerve supplies the *primary auditory area* along the side of the cerebrum. At first, these are just meaningless sounds. But the signals are sent out a bit

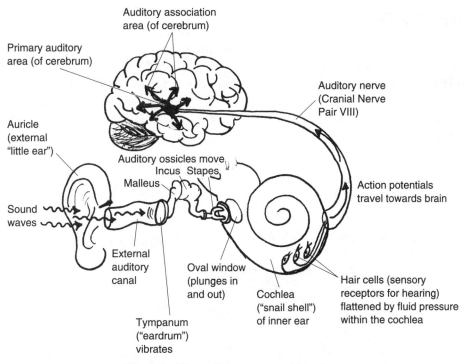

Fig. 9.7 The auditory nerve and hearing.

farther to reach the *auditory association area*. Here the meaningless jumbles of sounds are miraculously transformed into the recognizable words of understanding.

Other branches of Cranial Nerve VIII are involved in maintaining our upright posture and body balance.

Basic Parts of the Brain

We have been discussing the cerebrum indirectly, mostly as the place where various Special Senses from the cranial nerves are sent for processing. In general, we have the functional sequence: **Sensory Input–Integration or Association–Motor Output**, for the cerebrum.

We will now introduce major parts of the brain (starting with the cerebrum). And we will very briefly assign these brain parts a Sensory Input, Integration or Association, or Motor Output, functional classification.

Organ 1

THE CEREBRUM

The cerebrum (main brain mass) sits on top of the CNS like a big mushroom with two wrinkled "half-caps" called the *cerebral* (**seh-REE**-bral) *hemispheres* (**HEM**-ih-**sfeers**). The cerebrum is covered on its surface by the *cerebral cortex* (**KOR**-teks), a thin outer "bark" of *gray matter*. As Figure 9.8 shows, the cerebral cortex (outer bark of gray matter) is covered with numerous grooves or "furrows," termed *sulci* (**SUL**-see).

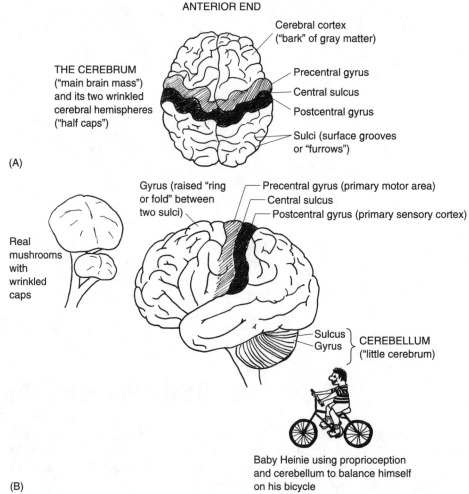

Fig. 9.8 The cerebrum and cerebellum. (A) Dorsal (overhead) view of the cerebrum (and its right and left cerebral hemispheres). (B) Lateral (side) view of cerebrum and the cerebellum.

Between each pair of sulci we see a *gyrus* (**JEYE**-rus). A gyrus is a raised "ring or fold" of brain tissue. Several *gyri* (**JEYE**-ree) are of particular functional importance. Consider the ones on either side of the *central sulcus*. As its name tells, the central sulcus is a groove that runs down the "center" of each cerebral hemisphere.

Just anterior or "before" (*pre-*) the central sulcus, is the *precentral gyrus*. The precentral gyrus is importantly associated with the motor output of the cerebrum. Specifically, it is called the *primary motor area* of the cerebral cortex. It is here that we consciously decide to move part of the body.

And just posterior or "after" (*post-*) the central sulcus, is the *postcentral gyrus*. The postcentral gyrus is closely tied to the sensory input of the cerebrum. Many of the *general body sensations* (such as touch, pressure, temperature, and pain), are eventually sent up into the postcentral gyrus for processing. This gives the postcentral gyrus the honor of being called the *primary sensory cortex*.

THE CEREBELLUM

The *cerebellum* (**sair**-uh-**BELL**-um) is the "little cerebrum" (*cerebell*) or the minor brain mass. It does seem, when looking at it, to be a miniature version of the cerebrum. The cerebellum, for example, like the cerebrum, has its surface covered with an extensive series of furrows (sulci) and rings or folds (gyri).

Organ 2

Located inferior and posterior to most of the cerebrum, the cerebellum is importantly involved in the unconscious coordination of posture, *reflexes*, and body movements. A reflex is an involuntary response to a particular "goad" or *stimulus* (**STIM**-you-lus). *Stimuli* (**STIM**-you-lie) for various reflexes involving the cerebellum are detected by *proprioceptors* (**proh**-pree-oh-**SEP**-tors). These are sensory "receptors" (*-ceptors*) that receive sensations from "one's own" (*propri*) self. By self, we mean the positions of your own self (body and body parts) in space.

The proprioceptors are largely found within the skeletal muscles and joints. Thus, you can sense or feel that you are standing upright, or that you have raised your arm off the table, without even having to look! Such proprioception (**proh**-pree-oh-**SEP**-shun) helps your body maintain its vertical balance and reduces the amount of shaking during automatic reflex movements.

It is the cerebellum that we train to help us successfully perform various *fine motor skills* (precise body movements) during our lifetimes. The cerebellum is progressively trained to help us maintain an upright posture while we walk, and to keep Baby Heinie from falling off his bicycle!

THE BRAINSTEM

If the cerebrum and cerebellum can be considered the spongy, wrinkled caps on top of a big mushroom, then the *brainstem* is the part that plugs into the bottom of the caps. The brainstem is the narrow, stemlike, inferior portion of the brain (Figure 9.9).

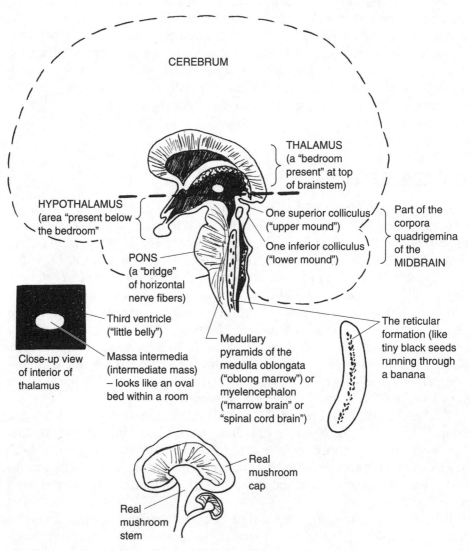

Fig. 9.9 Anatomy of the brainstem. Midsagittal section through interior of cerebrum and brainstem.

The thalamus

The top of the brainstem is called the *thalamus* (**THAL**-uh-**mus**). The thalamus is an egg-shaped "bedroom" (*thalam*). It looks like a room, because it contains most of the *third ventricle* (**VEN**-trih-**kl**) – a "little belly"-like cavity. The *massa* (**MAH**-sah) *intermedia* (in-ter-**ME**-dee-uh), also called the *intermediate mass*, is an oval mass within the thalamus. It looks somewhat like a round "bed." Hence,

THALAMUS = MASSA INTERMEDIA or + THIRD VENTRICLE
("Bedroom")   **INTERMEDIATE MASS**   ("Room"-like
               (Round "bed")   chamber)

The thalamus is often described as a *sensory relay center*. This is because it is a switching-over area for general sensory impulses (such as pain, touch, and temperature) on their way up to the cerebral cortex.

The hypothalamus

Just "below" (*hypo-*) the "bedroom" (*thalam*), is located the *hypothalamus*. The hypothalamus starts just below the level of the massa intermedia (intermediate mass). It contains a number of *control centers* for *homeostasis* (**hoh**-me-oh-**STAY**-sis). Homeostasis is a relative constancy of particular *anatomical or physiological parameters* (pah-**RAM**-eh-ters) within the body's internal environment. (Parameters are aspects of body structure or function which can be measured and expressed as a certain number of units.) For instance, the *temperature control center* for regulating the physiological parameter of oral body temperature in degrees Fahrenheit, is located in the hypothalamus. There is also a *hunger center*, *thirst center*, *pleasure center*, and many others.

The midbrain

Next in sequence below the hypothalamus, one finds the *midbrain*. As its name suggests, the midbrain is located in the "middle" of the brainstem, with two parts of the brainstem lying above it, and two lying below.

On the dorsal aspect of the midbrain are four prominent bumps called the *corpora* (**KOR**-por-ah) *quadrigemina* (**quad**-rih-**JEM**-ih-nah) or "four twin bodies." The two bigger bumps on top are the *superior colliculi* (kahl-**LIK**-you-lie) or "upper mounds." The superior colliculi help carry out *visual reflexes*, such as the automatic turning of the head and eyes to see some object (like a pesky fly) entering the *visual field*. The two smaller bumps on the bottom of the midbrain are the *inferior colliculi* (kahl-**LIK**-you-lie) or "lower mounds." The inferior colliculi are involved in carrying out *auditory reflexes*, such as the automatic turning of the head to point the ear in the direction of some sound.

The pons

Just below the midbrain is the *pons* (**PAHNS**). The pons is a large, beer belly-like bulge on the ventral aspect of the brainstem. The pons makes such a big bulge because it serves as a "bridge" of transverse nerve fibers that interconnects the two halves of the brainstem across the body midline.

The medulla oblongata or myelencephalon

The fifth and most inferior portion of the brainstem is the *medulla* (meh-**DUL-ah**) *oblongata* (**ahb**-long-**GAH**-tah). The phrase, medulla oblongata, exactly translates to mean "oblong marrow." The medulla is soft and cylinder-shaped (much like the yellow "marrow" in long bones), and it is "oblong" (significantly longer than it is wide). Its alternate name is *myelencephalon* (**my**-el-en-**SEF**-ah-lahn). This translates as either "marrow" (*myel*) "brain" (*cephalon*), or "spinal cord" (*myel*) "brain." This name fits, since the myelencephalon (medulla oblongata) is the bottom portion of the brain (and its brainstem), just above the spinal cord.

The medulla oblongata contains a number of *vital and nonvital reflex centers*. A reflex center, in general, is an organizing area for a reflex. Vital reflex centers are the organizing areas of reflexes necessary for "life" (*vit*). These include the *cardiac center*, which can both speed up and slow down the heart rate; as well as the *vasomotor* (**VAY**-soh-**moh**-ter) or "vessel" (*vaso-*) "movement" center, which causes blood vessels to either constrict (narrow) or dilate (widen), thereby influencing the blood pressure. Joining the other two vital reflex centers is the *respiratory* (**RES**-pir-ah-**tor**-ee) *center*, which automatically controls the rate and depth of breathing.

Nonvital reflex centers are those that control reflexes "not" (*non-*) necessary for life. These include the *gagging, coughing, sneezing, swallowing*, and *vomiting centers*.

Finally, we have the *medullary* (**MED**-you-**lair**-ee) *pyramids*. These are two long, pyramid-like swellings that run down the ventral aspect of the medulla oblongata. They bulge out because they contain thousands of motor nerve fibers that have descended from the cerebrum, and crossed over to direct movements on the opposite side of the body.

The reticular formation

Not forming a separate part of the brainstem, but scattered all the way through it, is the *reticular* (reh-**TIK**-you-lar) *formation*. The reticular formation is a "little network" (*reticul*) of tiny gray matter *nuclei* that are scattered up-and-down throughout the brainstem, much like the tiny black seeds in the center of a banana.

The reticular formation has an important role to play as the *Reticular Activating System* (*RAS*). You might think of the Reticular Activating System (RAS) as the body's own internal alarm clock. The RAS "activates" or "arouses" the cerebrum from sleep and helps it maintain a state of alert consciousness. It can wake you up early, even when you have the day off from work or school! The RAS likewise helps you tense your skeletal muscles, as when you wake up in the morning, and tense your muscles before you jump up out of bed.

BRAINSTEM SUMMARY

A

Let us now provide a quick summary equation for the 5 main parts of the brainstem:

BRAINSTEM = THALAMUS + HYPOTHALAMUS + MIDBRAIN + PONS + MEDULLA OBLONGATA

Organ 3

Spinal Cord and Meninges: The "Marrow" and Its "Mothers"

The medulla oblongata, you may remember, is the myelencephalon – "marrow brain" or "spinal cord" brain. The word root, *myel*, can therefore stand for either "marrow" or "spinal cord." Indeed, the *body* or main mass of the spinal cord is a rather long and thin "cord," much like the yellow marrow within a long bone. But, in this case, the spinal cord "marrow" lies inside of the *vertebral canal*, a fluid-filled cavity within the vertebral column or backbone (Figure 9.10).

There are 31 pairs of spinal nerves flanking the sides of the spinal cord body. These nerves bring sensory (afferent) information in, and take motor

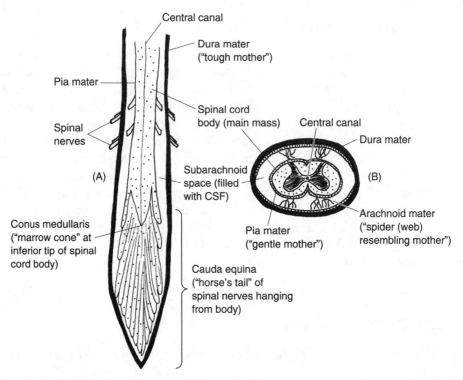

Fig. 9.10 The spinal cord and its neighbors. (A) The spinal cord: Longitudinal (lengthwise) view. (B) The spinal cord: Transverse (cross-sectional) view.

(efferent) information out, of the cord. Various *spinal reflex arcs* use the spinal cord and spinal nerves to help them carry out their automatic body functions. In addition, there are *ascending or sensory nerve tracts* that carry sensory information from the spinal cord "up" to higher levels of the CNS. Conversely, there are a number of *descending or motor nerve tracts* that carry motor (movement) information from the brain all the way "down" to the spinal cord.

Even though it contains lots of ascending and descending nerve tracts, the body of the spinal cord does not run the entire length of the vertebral column. Rather, it comes to a pointed tip called the *conus* (**KOH**-nus) *medullaris* (med-you-**LAIR**-is) or "marrow cone." The tip of the conus medullaris (marrow cone) ends between the *first and second lumbar vertebra* in most adults. Of the 31 pairs of spinal nerves, about 10 pairs hang down along either side of the conus medullaris to create a *cauda* (**KAW**-dah) *equina* (eh-**KWY**-nah) or "horse's tail."

THE CENTRAL CANAL AND CEREBROSPINAL FLUID (CSF)

Running down through the center of the spinal cord is the *central canal*. Like the third ventricle mentioned earlier (with the thalamus), the central canal is part of the *ventricular* (ven-**TRIK**-yew-lar) *system*. This system is a collection of "little bellies" (ventricles) that contain and circulate the *cerebrospinal* (ser-e-broh-**SPY**-nal) *fluid* or *CSF*.

The cerebrospinal fluid (CSF) is a clear, watery *filtrate* (**FIL**-trayt) – filtration product – of the blood *plasma* (**PLAZ**-mah). Somewhat like the blood plasma (liquid portion of the blood) circulates through the blood vessels, the CSF circulates through the cavities and passageways of the ventricular system. A major function of the cerebrospinal fluid is to serve as a shock absorber, protecting the delicate, butter-soft brain and spinal cord from knocking against the hard wall of surrounding bone tissue.

THREE "MOTHERS" PROTECT THE BRAIN AND CORD

A

Tissue 2

In Latin, your *mater* (**MAY**-ter) is your "mother"! An alternate name for your *maters* ("mothers") are the *meninges* (men-**IN**-jeez). Actually, the word, meninges, is more descriptive of the associated anatomy of the maters, because they are really flat, protective "membranes" that surround the CNS. Much like real "mothers," the three maters or meninges function to help protect the brain and spinal cord from physical trauma.

Each one of the maters (like our real mothers), has their own unique personalities:

1. Dura (**DUR**-ah) **mater** is the "tough mother"! It is the thick, tough, white outer membrane that lines the skull and vertebral canal. [**Study suggestion:** Visualize the brain and spinal cord as little peanuts within a shell. What does the red skin on the peanuts represent, anatomically speaking? – What does the shell represent?]

2. Arachnoid (ah-**RAK**-noyd) **mater** is the "spider-resembling mother." Rather than a spider, however, the arachnoid mater is more of a spider *web*-resembling mother! There is a delicate, spider web-like, branching network of collagen and elastic fibers extending "below" (*sub-*) the arachnoid mater, helping to create the *subarachnoid* (sub-ah-**RAK**-noyd) *space*. The subarachnoid space is filled with CSF, and it has some large blood vessels running through it.

3. Pia (**PEA**-uh) **mater** is literally the "soft or gentle" (*pia*) "mother." The pia mater gets its name from the fact that it dips down into the sulci (grooves) and gently hugs the surface of the brain and spinal cord. The pia mater is rich in tiny blood vessels that help deliver oxygen and other nutrients to the spinal cord tissue.

Spinal "Taps" Sample the "Sap"

Tissue 1

In summary, CSF circulates *through* the brain and spinal cord within the *ventricular system*, and *around the outside* of the brain and cord within the *subarachnoid space*. Because of its close, intimate contact with the actual neurons and nerve tissue, cerebrospinal fluid is often withdrawn and examined for possible abnormalities.

A *spinal tap* is usually called a *lumbar puncture*. The reason is that a long needle is inserted into the subarachnoid space between lumbar vertebrae #3 and 4, or 4 and 5. A small sample of CSF is withdrawn and examined for the presence of pus, blood, bacteria, cancer cells, or other problems. A quick diagnosis of an important Biological Disorder of the nervous tissue may therefore be obtained. [**Study suggestion:** Ask yourself: "Why isn't a spinal tap performed *above* the thoracic region of the vertebral column?"]

Quiz

Refer to the text in this chapter if necessary. A good score is at least 8 correct answers out of these 10 questions. The answers are listed in the back of this book.

1. The gross body soma is the:
    (a) Skin, joints, eyes, and smooth muscle
    (b) Big bag of guts (viscera)
    (c) Internal network of brain blood vessels
    (d) Big bag of skin, bones, joints, and skeletal muscles

2. The three primary germ layers:
    (a) Malleus, incus, and stapes
    (b) Arachnoid mater, pia mater, and dura mater
    (c) Superior, middle, and inferior colliculi
    (d) Mesoderm, endoderm, ectoderm

3. The Central Nervous System = ____ + ____:
    (a) Brain; medulla oblongata
    (b) Spinal cord; encephalon
    (c) Cerebrum; cerebellum

(d) Dura mater; CSF

4. Contains many control centers for homeostasis:
    (a) Hypothalamus
    (b) Body of spinal cord
    (c) Meninges
    (d) Superior colliculus

5. Somites in the embryo eventually create most of the:
    (a) CNS
    (b) Smooth and cardiac muscle
    (c) Soma
    (d) Psyche

6. The PNS = The ____ + ____:
    (a) Nerves; sensory receptors
    (b) Brain; spinal nerves
    (c) Spinal cord; spinal nerves
    (d) Thalamus; hypothalamus

7. The "self-regulating" portion of the Peripheral Nervous System:
    (a) Spinal effectors
    (b) Cerebral cortex
    (c) Autonomic Nervous System
    (d) "Special Senses"

8. Cranial nerve serving the sense of smell:
    (a) II, optic
    (b) I, olfactory
    (c) VII, facial
    (d) IX, glossopharyngeal

9. The major sensory relay area within the brainstem:
    (a) Stapes
    (b) Pons
    (c) Thalamus
    (d) Pyramids

10. Subconscious proprioception helping to coordinate fine body movements:
    (a) Retina
    (b) Cerebellum
    (c) Medulla oblongata
    (d) Fingernails

Body-Level Grids for Chapter 9

Several key body facts were tagged with numbered icons in the page margins of this chapter. Write a short summary of each of these key facts into a numbered cell or box within the appropriate *Body-Level Grid* that appears below.

***Anatomy* and *Biological Order* Fact Grids for Chapter 9:**

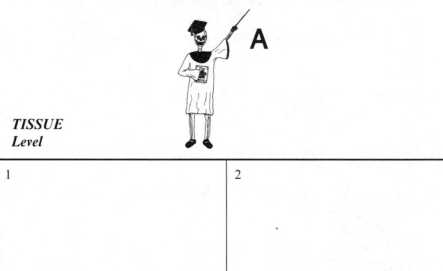

TISSUE
Level

1	2

ORGAN
Level

1	2
3	

ORGAN SYSTEM
Level

1	2
3	

ORGANISM
Level

1

Anatomy and *Biological Disorder* **Fact Grid for Chapter 9:**

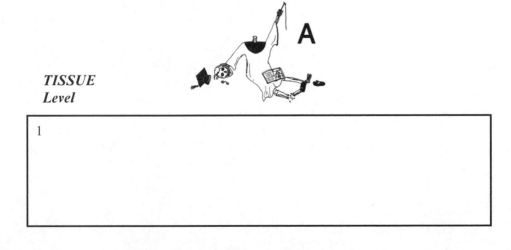

TISSUE
Level

1

Physiology and *Biological Order* Fact Grid for Chapter 9:

ORGAN
Level

1

Glands: The Secreting "Acorns"

In Chapter 9, we basically talked about the cellular "trees" – the neurons and their branching projections, the axons and dendrites. The Greek word, *dendrites*, does, in fact, literally "pertain to trees." Its many tree-like branches of cytoplasm allow a single neuron to, "Reach out and touch someone (almost)." A neuron's multiple branches may include axon terminals that stimulate muscle fibers at neuromuscular junctions (Figure 10.1, A). Neurons can also give or receive information from a network of many other neurons, via communication across synapses (Figure 10.1, B). Both neuromuscular junctions and synaptic junctions involve the same basic mode of cellular communication: the release of chemical neurotransmitter molecules (such as acetylcholine, ACh), which diffuse across a narrow, salt-water-filled gap.

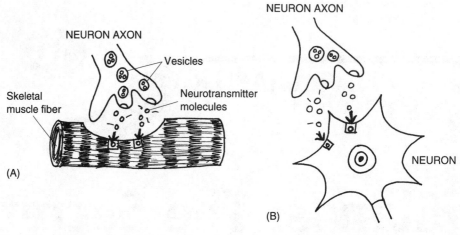

Fig. 10.1 Cellular communication by release of neurotransmitters (A) Neuromuscular junction. (B) Synapse.

Enter the "Acorns": The Glands Come Marching In!

We know something about "trees," don't we? Sometimes they "communicate" by dropping acorns! In the case of our human body, the "acorns" are actually *glands*. (These glands are often rounded, therefore shaped somewhat like "acorns.")

A gland is one or more epithelial cells specialized for the function of *secretion* – the release of some useful product. All living cells in the body, however carry out *excretion* (the release of waste products). The reason is that all living cells carry out metabolism, changing nutrients into energy (like ATP). However, since the conversion of potential (stored) energy within food into free (kinetic) energy is never 100% efficient, some waste products always result.

To summarize, let's state what we can call, the *Glandular* (**GLAN**-dyoo-lar) *Secretion Rule*:

Cell 1

> ***All*** living cells *excrete* waste products; but only *gland* cells *secrete* useful products.

An Overview of the Glandular System

Recall that the glandular system was first defined and introduced back in Chapter 3. Please go back and briefly review this material, now, so we won't have to keep repeating all of it. Especially, look back over the summary equation for the glandular system and Figure 3.3, which contrasted the *endocrine* glands with the *exocrine* glands. Have you reviewed it? Good!

THE PANCREAS – SOME GLANDS ARE JUST ALL *MIXED UP*!

Although most glands are either ductless (the endocrine glands) or have ducts (the exocrine glands), there is a third, hybrid category – the *mixed glands*. A mixed gland is part endocrine, plus part exocrine, both parts being mixed together within the same gland organ.

A key example of a mixed gland is the *pancreas* (**PAN**-kree-as). The word, pancreas, means "all" (*pan-*) "flesh" (*creas*). Certainly, a look at the pancreas will quickly reveal that it is, indeed, composed solely of soft, fleshy epithelial tissue. The pancreas resembles a comma, placed on its side, just along and behind the bottom edge of the stomach (Figure 10.2). And like the comma used to punctuate sentences, the pancreas has a *head*, main *body*, and a *tail*. The head of the pancreas is attached to the side of the *duodenum* (dew-**AHD**-eh-num), which is the first or proximal portion of the *small intestine*.

Running down the middle of the pancreas is the hollow *pancreatic* (pan-kree-**AT**-ik) *duct*. The pancreatic duct has numerous small branches. Around the hollow tips of the smallest branches are the *pancreatic acini* (**AS**-uh-neye) – clusters of gland cells arranged like "little grapes" (*acin*). The pancreatic *acinar* (**AS**-uh-nar) *cells* secrete the *pancreatic* (pan-kree-**AT**-ik) *juice*. This juice is a mixture of various digestive enzymes and other materials, which enter the pancreatic duct. The pancreatic acini and the pancreatic duct, therefore, represent the exocrine (duct-bearing) portion of the pancreas.

Scattered here and there within the pancreas are a bunch of *islets* (**I**-lets) or "little islands" of endocrine gland cells. Technically speaking, these are called the *pancreatic islets* ("little islands of the pancreas") or the *Islets of Langerhans* (**LAHNG**-ur-hahns). These little islands of endocrine gland cells get their name from Paul *Langerhans* (**LAHNG**-ur-hahns), a German physician and anatomist who first described them.

There are two main epithelial cell populations within each pancreatic islet (Islet of Langerhans). These are called the *alpha* (**AL**-fuh) *cells* and the *beta*

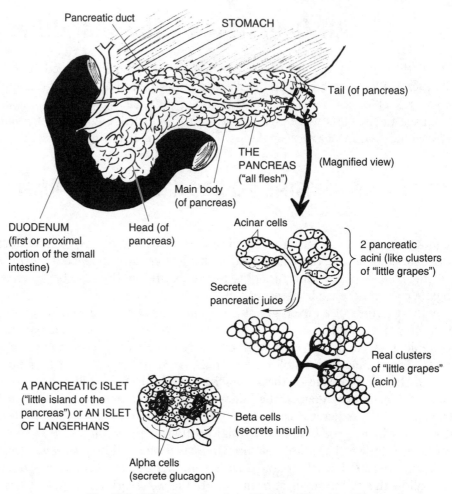

Fig. 10.2   The pancreas – A really "mixed"-up gland!

(**BAY**-tuh) *cells*. The beta cells secrete the critical hormone, *insulin* (**IN**-suh-lin). This hormone is well-known for its effect of *lowering* the blood glucose concentration. The much-less-recognized alpha cells, in contrast, secrete a hormone called *glucagon* (**GLOO**-kah-gahn). Directly opposite to insulin, glucagon *raises* the blood glucose concentration.

In summary of the pancreatic islets (Islets of Langerhans):

ALPHA CELLS ⟶ Secrete *glucagon* into bloodstream ⟶ *Raises* **the blood glucose concentration**

BETA CELLS $\longrightarrow$ Secrete *insulin* into bloodstream $\longrightarrow$ *Lowers* the blood glucose concentration

A

GLANDULAR SYSTEM SUMMARY

Let us now provide a final summary equation for the glandular system:

Organ System 1

GLANDULAR = SYSTEM	ENDOCRINE GLANDS *(Internal secretion of hormones – chemical messengers – directly into the bloodstream)*	+	EXOCRINE GLANDS *(External secretion of useful products into ducts)*	+	MIXED GLANDS *(Contain both endocrine and exocrine components)*

The Major Endocrine Glands and Their Hormones

Since they are of such great importance, this section will focus upon the major endocrine glands and their hormones:

Endocrine gland #1: The pancreatic islets (Islets of Langerhans). (This gland and its hormones, of course, were discussed in the previous section.)

Endocrine gland #2: The thyroid gland. Along with the pancreas, perhaps the most-recognized endocrine gland is the *thyroid* (**THIGH**-royd) *gland*. The thyroid gland literally "resembles" (*-oid*) two large, oblong "shields" (*thyr*) – as from a pair of African warriors pressing hands under the chin (Figure 10.3). The thyroid gland busily extracts the element *iodine* from the bloodstream, then incorporates it into its most important hormone, *thyroxine* (thigh-**RAHKS**-in). Thyroxine stimulates mitosis (cell division) and protein synthesis, thereby promoting body growth and tissue repair. It also tends to raise the *BMR* or *Basal* (**BAY**-sal) *Metabolic Rate*. The BMR (Basal Metabolic Rate) is the rate at which the body cells operate when they are under "resting" or "basal" conditions. In general, this means the times when we are neither exercising nor digesting any food. Thus, because it speeds up the metabolism, thyroxine tends to raise body temperature, as well.

Endocrine gland #3: The adrenal cortex – outer "bark" of the adrenal body. There is a double-gland lying just "above" (*supra-*) each "kidney" (*renal*).

F

Molecule 1

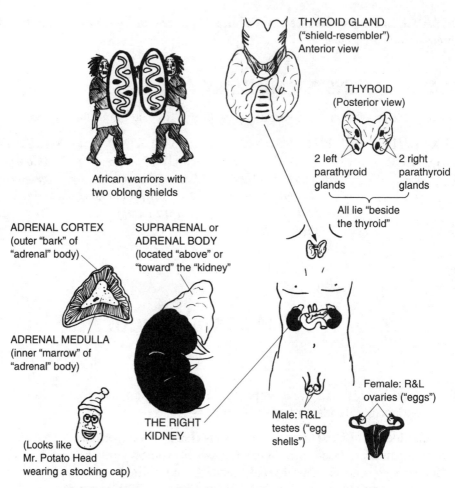

Fig. 10.3   Major endocrine glands of the neck and trunk regions.

Alternatively, it can be described as being positioned "toward" (*ad*-) each "kidney" (*renal*). Putting the preceding information together, we obtain two different names – *suprarenal* (**soo**-prah-**REE**-nal) *body* or *adrenal* (uh-**DREE**-nal) *body*. The suprarenal (adrenal) body is a double-gland that sits atop each kidney, much like a pointed cap perched on top of Mr. Potato Head (the kidney).

The double-gland first encountered is the one on the outside, called the *adrenal cortex* (**KOR**-teks). The adrenal cortex is the thin, outermost "bark" of the adrenal body. The adrenal cortex secretes two major groups of steroid hormones: the *glucocorticoids* (**gloo**-koh-**KORT**-ih-koyds) and the *mineralocorticoids* (**min**-er-al-oh-**KOR**-tih-koyds).

The glucocorticoids are literally "steroids" (*-oids*) from the adrenal "cortex" (*cortic*) that affect blood "glucose" (*gluc*) levels. The major hormone in the glucocorticoid family is called *cortisol* (**KOR**-tih-sahl). Cortisol tends to protect our bodies against stress, reduce the symptoms of tissue inflammation, and raise blood glucose levels. [**Study suggestion:** What hormone have we already studied that also acts to raise blood glucose concentration?]

Molecule 2

The mineralocorticoids are "steroids" (*-oids*) from the adrenal "cortex" (*cortic*) that affect blood "mineral" (*mineral*) levels. In particular, the mineralocorticoid family includes the specific hormone called *aldosterone* (al-**DAHS**-ter-ohn). The most critical physiological effect of aldosterone is raising the blood concentration of sodium (Na^+) ion, a very significant body mineral.

Molecule 3

Endocrine gland #4: The adrenal medulla – inner "marrow" of the adrenal body. We said that the adrenal (suprarenal) body was a double-gland, didn't we? The outer part is the more peripheral endocrine gland, the adrenal cortex. Another, completely separate endocrine gland, hidden within, is the *adrenal medulla* (muh-**DULL**-ah).

The adrenal medulla is the inner "marrow" of the adrenal (suprarenal) body. This endocrine gland specializes in the secretion of two *catecholamines* (**cat**-uh-**KOHL**-uh-means). The last part of the catecholamine name is *amine*, indicating that these hormones are derived from an *amino* acid, called *tyrosine* (**TIE**-roh-**seen**). *Epinephrine* (**ep**-ih-**NEF**-rin) and *norepinephrine* (nor-**ep**-ih-**NEF**-rin) are the two principal catecholamines secreted by the adrenal medulla. Their alternate names are *adrenaline* (ah-**DREN**-uh-lin) for epinephrine and *noradrenaline* (**nor**-ah-**DREN**-uh-lin) for norepinephrine.

Epinephrine (adrenaline) and norepinephrine (noradrenaline) are very central in promoting the Emergency "Fight-or-Flight" Response to severe stress. When a person feels threatened or just "stressed-out," symptoms of the "Fight-or-Flight" Response develop. You probably have experienced, for instance, feelings of a much faster pulse, faster and deeper breathing, and your heart pounding harder within your chest. These physiological reactions to stress are closely related to the activity of the sympathetic portion of the Autonomic Nervous System, and to the effects of epinephrine and norepinephrine upon their visceral target organs. (Review these topics in Chapter 9, if desired.)

Molecule 4

Endocrine gland (group) #5: The four parathyroid glands. Lying "beside" (*para-*), and just in back of the "thyroid" gland, are the four small *parathyroid* (per-uh-**THIGH**-royd) *glands*. These glands secrete *parathyroid* (per-uh-**THIGH**-royd) *hormone*, abbreviated as *PTH*. The four parathyroids are stimulated to secrete more PTH whenever the blood calcium concentration begins to fall below its average or "set-point" level. The PTH travels to the bone matrix within the skeleton, where it increases the *resorption* (re-**SORP**-shun) or breaking down of the calcium phosphate crystals covering bone collagen fibers. When these calcium-containing bone crystals are partially broken down, their stored calcium (Ca^{++}) ions are released into the blood-stream. This physiological effect thus helps raise the blood calcium ion concentration back towards its average, long-term levels.

Endocrine gland #6: The ovaries in the female. In fertile females, there is a pair of *ovaries* (**OH**-var-eez) – whitish, oval endocrine glands that look quite a bit like "eggs" (*ov*). The process of *oogenesis* (oh-oh-**JEN**-uh-sis), the "formation of" (*genesis*) a mature *ovum* (**OH**-vum) or "egg" cell, goes on within the two ovaries.

Molecule 5

Two major hormones are secreted during various stages of oogenesis. These hormones are *estrogen* (**ES**-troh-jen) and *progesterone* (proh-**JES**-ter-ohn). Estrogen is primarily responsible for the development of the *secondary sex characteristics* in the female. The secondary sex characteristics are those traits that have nothing directly to do with reproduction or childbirth. They include soft skin, higher voice, less body hair, and a greater proportion of body fat, compared to males. Progesterone prepares the female body for pregnancy, as, for example, by stimulating the thickening of the lining of the *uterus* (**YEW**-ter-us) or "womb." This makes the *uterine* (**YEW**-ter-in) lining more plush and glandular, just in case a fertilized ovum is implanted into it.

Endocrine gland #7: The testes in the male. Males have a pair of *testes* (**TES**-teez) – whitish, oval endocrine glands – located within the *scrotum* (**SKROH**-tum), a hairy external bag of skin. Just as the ovaries are involved in oogenesis, the testes are involved in *spermatogenesis* (sper-mat-uh-**JEN**-eh-sis). Spermatogenesis is the "production of" (*-genesis*) "sperm" (*spermat*) cells.

Molecule 6

And, like the ovaries, the testes also function as endocrine glands. But they secrete *testosterone* (tes-**TAHS**-ter-ohn). Testosterone stimulates the formation of the *secondary sex characteristics* in the male. About the opposite of estrogen's effects, testosterone results in thicker skin and bones, deeper voice, greater amounts of body hair, and a greater proportion of skeletal muscle tissue, compared to females. It also stimulates the sex drive.

Endocrine gland #8: The anterior pituitary gland. We talked about the brain in Chapter 9. Recall that the brain includes two neighboring regions, the thalamus and hypothalamus. (Review Figure 9.9, if desired.) Located just below the hypothalamus, we find the *pituitary* (pih-**TOO**-ih-**tair**-ee) *body*. The pituitary body is attached to the underbelly of the hypothalamus, by means of a hollow *pituitary* (pih-**TOO**-ih-**tair**-ee) *stalk* (see Figure 10.4, A).

One often sees reference to the *pituitary gland*. But, technically, there is a small, gray, round, pituitary *body* that contains at least two separate

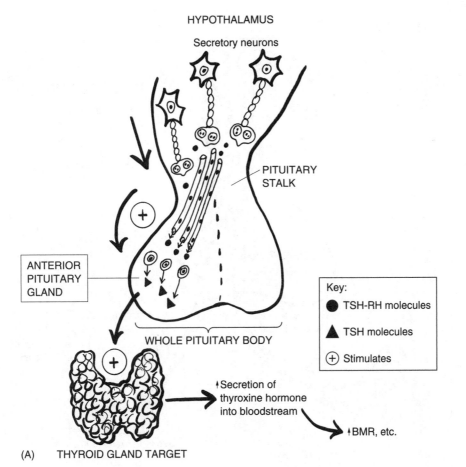

HYPOTHALAMUS

Secretory neurons

PITUITARY
STALK

ANTERIOR
PITUITARY
GLAND

Key:

● TSH-RH molecules

▲ TSH molecules

⊕ Stimulates

WHOLE PITUITARY BODY

↕Secretion of
thyroxine hormone
into bloodstream

↕BMR, etc.

(A)   THYROID GLAND TARGET

Fig. 10.4   The hypothalamus and pituitary body. (A) The hypothalamus and the anterior
pituitary gland. (B) The hypothalamus and the posterior pituitary gland. ADH =
Antidiuretic hormone. BMR = Basal metabolic rate. TSH = Thyroid-stimulating
hormone. TSH-RH = Thyroid-stimulating hormone releasing hormone.

endocrine glands. The one located more in front of the pituitary body is
appropriately named the *anterior pituitary gland.*

The anterior pituitary gland secretes a group of hormones called the
trophic (**TROHF**-ik) or "nourishing" hormones. These trophic hormones
have as their main targets or effectors other endocrine glands, which they
"nourish" (stimulate) to secrete their own individual hormones. Consider, for
example, *thyroid-stimulating hormone,* which is abbreviated as *TSH.* As its
name reveals, thyroid-stimulating hormone "nourishes" or stimulates the
thyroid gland to increase the secretion of its main hormone, thyroxine.

F

Molecule 7

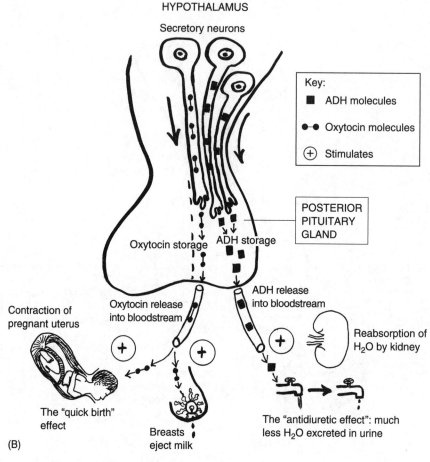

Fig. 10.4 (continued)

Endocrine gland #9: The posterior pituitary gland. Located just behind the anterior pituitary gland, is the *posterior pituitary gland*. The posterior pituitary is contained within the pituitary body, like the anterior pituitary, but it is a completely different gland, and it is involved with different hormones.

Chief among these is *antidiuretic* (**an**-tih-die-you-**RET**-ik) *hormone*, which is abbreviated as *ADH*. A second hormone associated with the posterior pituitary gland is called *oxytocin* (**ahk**-see-**TOH**-sin) (see Figure 10.4, B).

Antidiuretic hormone literally acts "against" (*anti-*) "diuresis" (**die**-yew-**REE**-sis). Diuresis, in turn, comes from the Greek for "urinate." Thus, diuresis is the excretion of large amounts of urine. Antidiuretic hormone (ADH) circulates through the bloodstream, to the kidneys. Here it increases the *reabsorption* (**ree**-ab-**SORP**-shun) of water from the interior of the *kidney tubules* (**TWO**-byools) or "tiny tubes." More H$_2$O molecules leave the kidney tubules, and return to the bloodstream. This process reduces the amount of water in the kidney tubules, hence the amount of water excreted within the urine. (This is the meaning of an "antidiuretic" effect.)

The other hormone linked to the posterior pituitary is oxytocin. It is labeled for its ability to stimulate the contractions of the female uterus and thereby provide "quick" (*oxy-*) "birth" (*toc*) during labor. It also stimulates the breasts to eject milk for nursing.

Endocrine gland #10: The hypothalamus. Interestingly enough, the hypothalamus secretes separate hormones that affect both the anterior and the posterior pituitary glands (see Figure 10.4, A and B). Oxytocin and ADH are really secreted by modified nerve cells called *secretory* (**SEE**-kreh-**tor**-ee) *neurons*. The role of the posterior pituitary gland, therefore, is just to temporarily store, then release, both of these hormones into the bloodstream. They are actually a product of the hypothalamus, however.

Other secretory neurons in the hypothalamus produce *Releasing Hormones*, *RHs*. Such Releasing Hormones (RHs) circulate from the hypothalamus, through the pituitary stalk, and down into the anterior pituitary gland. Here they stimulate the *release* of particular trophic hormones. One of these can be called *Thyroid-Stimulating Hormone Releasing Hormone*, or *TSH-RH*. The TSH-RH molecules circulate through the pituitary stalk and stimulate the anterior pituitary gland to produce more TSH (Thyroid-Stimulating Hormone). And, as we have said earlier, the TSH, in turn, stimulates the thyroid gland to manufacture more thyroxine. This increases the BMR, and has other effects, such as enhancing growth, cell division, and protein synthesis.

Molecule 8

Pathological Anatomy of Glands: "What Happens When an Acorn Falls Off?"

If the glands are collectively known in Latin as "acorns," then what happens when one of these glandular "acorns" falls off the "tree" (human body)? Less poetically, we are talking about the issue of *gland ablation* (ab-**LAY**-shun). Any type of *ablation* (ab-**LAY**-shun), whether it be of a gland or some other body part, is a "taking away" or removal. A gland ablation, as a consequence, is a complete or partial removal of a gland.

Sometimes a gland must be *ablated* (ab-**LAYT**-ed) or "taken away" by surgery. One possible reason for such ablation might be that the gland was destroyed by some terrible accident to the body (auto accident, gunshot wound, and the like). Another reason is gland ablation due to the presence of a *malignant* (mah-**LIG**-nant) or "deadly" tumor. *Cancer of the thyroid gland*, for instance, might make *thyroidectomy* (thigh-royd-**EK**-toh-mee) – the "removal of" (-*ectomy*) the "thyroid" – absolutely necessary for the patient's survival.

Organ 1

In such cases of gland ablation or removal, there is likely to be a severe *hyposecretion* (**HIGH**-poh-see-**KREE**-shun) of the gland's hormones. Hyposecretion is a "deficient or below normal" (*hypo-*) "secretion" of some substance. After thyroidectomy (ablation or removal of the thyroid), for example, the patient may suffer from a hyposecretion of thyroxine. The resulting disease state is often called *hypothyroidism* (**high**-poh-**THIGH**-royd-izm).

The patient with hypothyroidism will tend to have an extremely low BMR (basal metabolic rate), an abnormally low body temperature, sleep frequently, and be mentally unresponsive and sluggish. The major cause of these symptoms, of course, is hyposecretion (or nearly absent) secretion of thyroxine into the bloodstream.

One common clinical solution is *hormone replacement therapy*. The physician may prescribe various synthetic hormones, or hormones from animals or other natural sources, to partially replace the hormones lost due to gland ablation.

Quiz

Refer to the text in this chapter if necessary. A good score is at least 8 correct answers out of these 10 questions. The answers are listed in the back of this book.

1. All body cells have excretions, but only gland cells have:
   (a) High BMR
   (b) Enzymes for metabolism
   (c) Secretion
   (d) Digestion

2. A gland containing both endocrine and exocrine components:
   (a) Mixed
   (b) Ductless

    (c) Sweat

    (d) Ducted

3. Grape-like clusters of gland cells that secrete pancreatic juice:
    (a) Alpha
    (b) Beta
    (c) Acinar
    (d) Islets

4. Secrete glucagon into the bloodstream:
    (a) Pancreatic acini
    (b) Fred's follicles
    (c) Alpha cells
    (d) Delta triangles

5. A major physiological effect of thyroxine:
    (a) Increased BMR
    (b) Storage of Ca^{++} ions within bone matrix
    (c) Enhanced depth of sleep
    (d) Lowered body temperature

6. Chief source of the body's epinephrine and norepinephrine:
    (a) Parathyroid glands
    (b) Adrenal medulla
    (c) Ovaries
    (d) Adrenal cortex

7. Stimulates formation of secondary sex characteristics in the male:
    (a) Progesterone
    (b) Oxytocin
    (c) Melatonin
    (d) Testosterone

8. An important anterior pituitary trophic hormone:
    (a) TSH
    (b) ADH
    (c) Insulin
    (d) Estrogen

9. Stimulates the walls of the uterus to contract during labor:
    (a) Parathyroid hormone
    (b) H_2O
    (c) Mucus
    (d) Oxytocin

10. Responsible for production of Releasing Hormones (RHs):
    (a) Hypothalamus
    (b) Midbrain
    (c) Thalamus
    (d) Posterior pituitary gland

Body-Level Grids for Chapter 10

Several key body facts were tagged with numbered icons in the page margins of this chapter. Write a short summary of each of these key facts into a numbered cell or box within the appropriate *Body-Level Grid* that appears below.

Anatomy and *Biological Order* **Fact Grid for Chapter 10:**

A

ORGAN SYSTEM
Level

1

Physiology and *Biological Order* Fact Grid for Chapter 10:

CELL
Level

1

Function and *Biological Order* Fact Grids for Chapter 10:

MOLECULE
Level

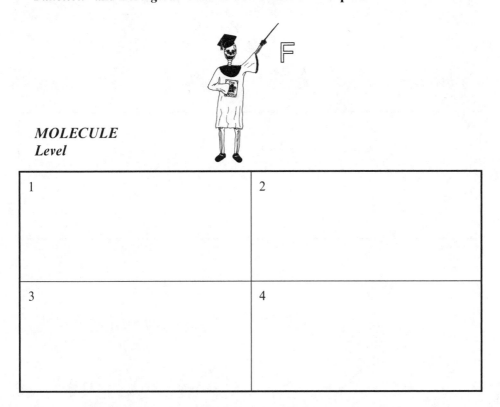

1	2
3	4

5	6
7	8

Physiology and *Biological Disorder* Fact Grids for Chapter 10:

ORGAN
Level

1

Test: Part 4

DO NOT REFER TO THE TEXT WHEN TAKING THIS TEST. A good score is at least 18 (out of 25 questions) correct. Answers are in the back of the book. It's best to have a friend check your score the first time, so you won't memorize the answers if you want to take the test again.

1. Psyche versus Soma means:
    (a) "Mind versus Body"
    (b) "Nerve versus Gland"
    (c) "Muscle versus Bone"
    (d) "Skeleton versus Brain"
    (e) "Brain versus Brawn"

2. The "little stomach" with the three primary germ layers in the embryo:
    (a) Gross soma
    (b) Little puppeteer
    (c) Germinativum
    (d) Stratum corneum
    (e) Gastrula

3. Sensory nerve fibers are also classified as:
   (a) Efferent
   (b) Motor
   (c) Afferent
   (d) Stimuli
   (e) Axons

4. The "self-regulating" portion of the PNS:
   (a) Somatic Nervous System
   (b) CNS
   (c) Autonomic Nervous System
   (d) Endocrine glands
   (e) Brain

5. Includes input from the "Special Senses":
   (a) Spinal cord
   (b) Visceral target organs
   (c) Smooth muscle effectors
   (d) SNS
   (e) Guts

6. By "Dual Innervation" of the viscera, it is meant that:
   (a) The brain receives more than one set of nerves
   (b) Internal organs are supplied by both sympathetic and parasympathetic nerves
   (c) Our guts can be both relaxed and excited at the same time
   (d) Parasympathetic nerves always act like the brake pedal on a car
   (e) Sympathetic nerves always act like the gas pedal on a car

7. Cranial nerve supplying our sense of vision:
   (a) Number I
   (b) Optic
   (c) Oculomotor
   (d) Number X
   (e) Auditory

8. Helps with the sense of hearing:
   (a) Tympanum
   (b) Retina
   (c) Rods
   (d) Cones
   (e) Posterior chamber

9. The thin "bark" of gray matter covering the main brain mass:
   (a) Medulla
   (b) Thalamus
   (c) Massa intermedia
   (d) Cerebral cortex
   (e) Horns of gray matter

10. The ____ is an egg-shaped "bedroom" at the top of the brainstem:
    (a) Cerebellum
    (b) Midbrain
    (c) Superior colliculus
    (d) Spinal cord body
    (e) Thalamus

11. Alternately called the myelencephalon or "spinal cord brain":
    (a) Medulla oblongata
    (b) Pons
    (c) Corpora quadrigemina
    (d) Hypothalamus
    (e) Stapes

12. Serves as the body's internal alarm clock:
    (a) Inferior colliculi
    (b) Stretch reflex
    (c) Vertebral canal
    (d) Brachial plexus
    (e) Reticular Activating System

13. The pointed inferior tip of the spinal cord:
    (a) Ventricular system
    (b) Descending nerve tracts
    (c) Pons
    (d) Conus medullaris
    (e) Dorsal root

14. The "tough mother" lining the skull and vertebral canal:
    (a) Arachnoid mater
    (b) Subarachnoid space
    (c) Dura mater
    (d) Central sulcus
    (e) Pia mater

15. Both the neuromuscular junction and synapse involve cell-to-cell communication by:
    (a) Release of chemical neurotransmitter molecules
    (b) Action potential waves which "hop" across the gaps
    (c) Glue-like cells that directly link the cells together
    (d) Binding of hormone molecules to their membrane receptor sites
    (e) Direct transmission of electrical radio signals

16. A good example of a mixed gland:
    (a) Ovary
    (b) Pancreas
    (c) Testis
    (d) Hypothalamus
    (e) Anterior pituitary

17. Gland of external secretion into ducts:
    (a) Exocrine
    (b) Adrenal cortex
    (c) Endocrine
    (d) Thalamus
    (e) Pineal

18. The major source of cortisol and other glucocorticoid hormones:
    (a) External auditory canal
    (b) Pancreatic juice
    (c) Adrenal medulla
    (d) Parathyroids
    (e) Adrenal cortex

19. Epinephrine is alternately known as:
    (a) Noradrenaline
    (b) Acetylcholine
    (c) Adrenaline
    (d) Thyroxine
    (e) PTH

20. Increases the resorption (breaking down) of bone matrix within bones:
    (a) Oxytocin
    (b) Parathyroid hormone
    (c) Progesterone
    (d) Wheaties
    (e) BMR

21. Estrogen helps produce:
    (a) Mature sperm cells
    (b) A "bridge" across the brainstem
    (c) Two gastrulas
    (d) Secondary sex characteristics in women
    (e) Testosterone secretion

22. Source of the so-called trophic hormones:
    (a) Thyroid
    (b) Ovaries
    (c) Posterior pituitary
    (d) Pituitary stalk
    (e) Anterior pituitary

23. Increases the reabsorption of water from kidney tubules:
    (a) Aldosterone
    (b) Antidiuretic hormone
    (c) TSH-RH
    (d) Glucagon
    (e) Insulin

24. The essential role played by the posterior pituitary gland:
    (a) Storage and release of two important hormones
    (b) Synthesis and secretion of all brain hormones
    (c) Secretion of oxytocin; storage of ADH
    (d) Manufacture of insulin
    (e) Heavy consumption of blood glucose

25. Endocrine gland ablation (removal) would most probably cause:
    (a) Bloody stools
    (b) Hypersecretion of the gland's hormones
    (c) Excessive urination and thirst
    (d) Hyposecretion of the gland's hormones
    (e) Soul-searching and regrets

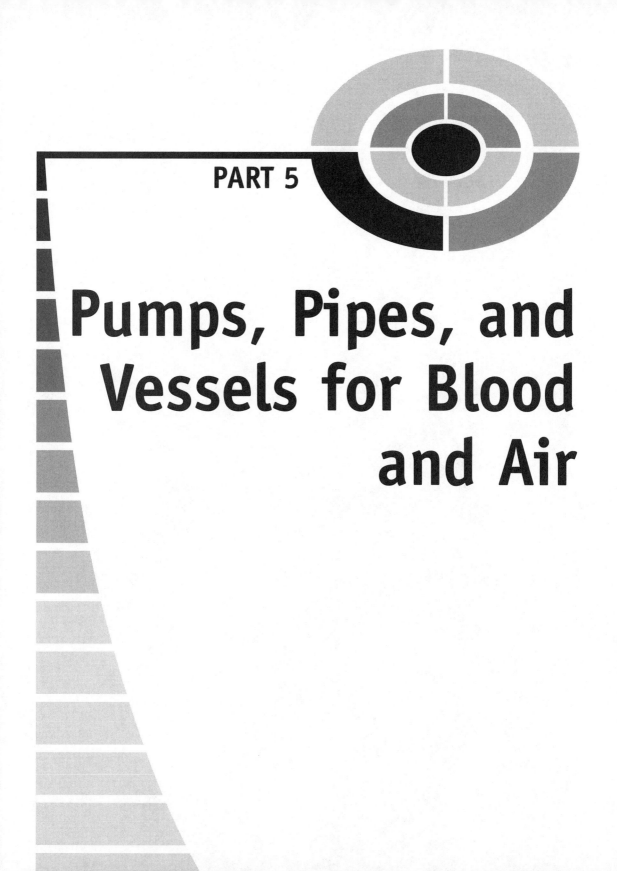

PART 5

Pumps, Pipes, and Vessels for Blood and Air

CHAPTER 11

The Circulatory System: A *Vasc* Responsibility To Play Our "Cardi's" Right!

The last couple of chapters have discussed the anatomy of the nerves and glands. There was also mention of the fact that both glands and muscle fibers are body *effectors*. This means that they carry out a particular function, thereby having some significant physiological *effect*.

The Heart and Vascular Network of the Circulatory System

In this chapter, we will include a discussion of muscular effectors that are involved with the *circulatory* or *cardiovascular system*. Chapter 3 provided an overview of this organ system, along with a summary equation for it. [**Study suggestion:** Go back to Chapter 3 and take a quick look at its brief overview of the circulatory system.] We will now re-sketch this overview in considerably more detail, so that it includes the names of the major types of blood vessels (see Figure 11.1).

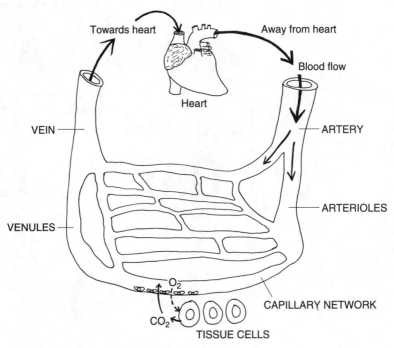

Fig. 11.1   The heart and vascular network of the circulatory system.

Organ System 1

The summary equation from Chapter 3 is also considerably expanded to read as follows:

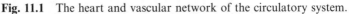

CIRCULATORY (CARDIOVASCULAR) SYSTEM	=	THE HEART ORGAN	+	VASCULAR NETWORK	+	BLOOD CONNECTIVE TISSUE

This organ system traces a "little circle" (*circul*) that both begins and ends with the same organ – the "heart" or *cardium* (**KAR**-dee-um). The circular pattern is created by the "little vessels" (*vascul*), which first leave the heart, then curve back and eventually return to it – boomerang, anyone? This very impressive, Biologically Ordered anatomy can be called a *vascular* (**VAS**-kyoo-lar) *network*. It does "involve or pertain to" (*-ar*) a bunch of interconnected "little vessels" (*vascul*), doesn't it?

GENERAL COMPONENTS OF THE VASCULAR NETWORK

Following what we see in Figure 11.1, the vascular network includes the *arteries*, *arterioles* (ar-**TEER**-ee-ohls), *capillary* (**CAP**-ih-**lair**-ee) *network*, *venules* (**VEN**-yools), and finally, the *veins*.

Summarizing these components, we obtain:

THE VASCULAR = ARTERIES + ARTERIOLES + CAPILLARY
(BLOOD VESSEL)   NETWORK + VENULES + VEINS
NETWORK

A

Organ 1

Finally, *blood connective tissue* is the material which is pumped in a "little circle" through the vascular network. As the blood circulates, it provides nutrients to the cells of most body tissues, and then picks up metabolic waste products from them. The force for pumping the blood comes, of course, from the contracting muscle tissue within the walls of the heart.

Cardiac and Smooth Muscle as Visceral Effectors

Chapter 9, you may remember, contrasted the Somatic Nervous System (SNS) with the Autonomic Nervous System (ANS). The SNS serves the gross body soma, which mainly features the skin, bones, joints, and skeletal muscle fibers as *somatic effectors*. Brief mention was also made about the ANS serving the *visceral target organs* or *"gut" effectors*. These visceral effectors include smooth muscle, cardiac muscle, and glands.

Well, Chapter 10 covered the glandular type of visceral effectors. Now, isn't it high time that we begin discussing the *other* two types of visceral effectors within the walls of our "guts" (viscera or internal organs) – *cardiac muscle tissue* and *smooth muscle tissue*? The location of cardiac muscle tissue

leads us to the heart, while the search for smooth muscle tissue guides us to the major blood vessels.

The Heart and Its Wall

Back in Chapter 5, we saw a picture of skeletal muscle tissue, located within the calf muscle or *gastrocnemius* (**gas**-trahk-**NEE**-me-us). This same tissue type can also be found in all of the over 600 individual skeletal muscle organs. You may recollect that these skeletal muscle organs are like a bunch of "little mice."

LOCATION OF THE HEART

Cardiac muscle tissue, in great contrast, has only one home in the entire body – the wall of the "heart" (*cardi*). And you have to dig down through several *gross membranes* around the heart, before you can even see it! The heart organ, itself, is a fist-sized double-pump located within the *mediastinum* (**me**-de-as-**TIE**-num).

As Figure 11.2 clearly illustrates, the mediastinum is the "middle" (*medi*) portion of the *thoracic* (thor-**ASS**-ik) or "chest" *cavity*. As such, it lies immediately deep to the breastplate or sternum (*stin*). It is flanked on either side by the right and left lungs.

It is helpful, anatomically, to visualize the heart as an inverted (upside-down) triangle. The *cardiac base* is thus the broad, flat, superior region of the heart. It is the place where most of the major blood vessels attach to the heart. Conversely, the *cardiac apex* (**AY**-peks) is the pointed, inferior "tip" of the heart.

In most people, the cardiac apex points distinctly to the left. Hence, even though the heart is situated within the mediastinum, about 2/3 of the heart is present on the *left* side of the body midline, rather than being exactly centered *on* the midline.

THE HEART WALL

The heart is enclosed within a tough double-sac, called the *pericardium* (per-uh-**KAR**-dee-um). The pericardium is literally "present" (-*um*) "around" (*peri-*) the "heart" (*cardi*). The outer portion of this double-sac is called the *parietal pericardium*, since it is the thin membrane closest to the "wall"

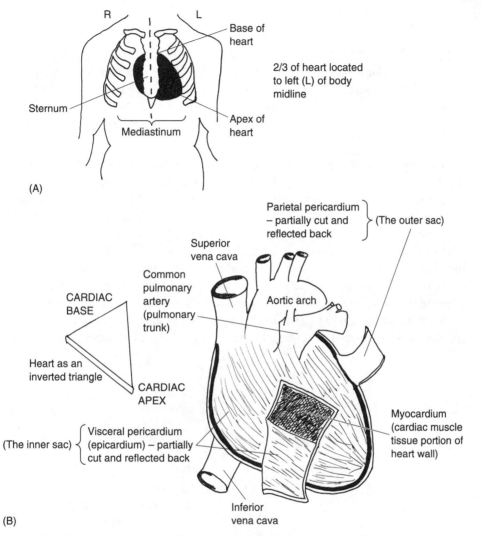

Fig. 11.2 The heart and its wall. (A) Location of the heart within the mediastinum. (B) External view of heart surface, showing pericardium (parietal & visceral).

(*parietal*) of the thoracic cavity. But the inner portion of this double-sac is called the *visceral pericardium*, since it is the thin membrane closest to the heart, itself, which is a type of "gut" or "internal organ" (*viscera*).

The visceral pericardium is alternately called the *epicardium* (**ep**-uh-**KAR**-dee-um). This is because the visceral pericardium is a thin, milky membrane that lies directly "upon" (*epi-*) the surface of the "heart" (*cardium*).

Tissue 1

Collecting this information yields:

THE PERICARDIUM =	**PARIETAL** +	**VISCERAL**
(*A tough double-sac "around heart"*)	**PERICARDIUM** (*Outer sac near chest "wall"*)	**PERICARDIUM (EPICARDIUM)** (*Inner sac "upon heart" or "gut" surface*)

If we *reflect* (cut and "bend back") the milky visceral pericardium (epicardium), the main portion of the heart wall is revealed. It is called the *myocardium* (**my**-oh-**KAR**-dee-um). The myocardium is the thick, middle, "cardiac muscle" (*myo*) tissue portion of the heart wall. A close look at the myocardium (as with a microscope) reveals that it consists of millions of *cardiac muscle fibers*. These are actually long, thin, fiber-shaped, cardiac muscle cells. And, like the skeletal muscle fibers, the cardiac muscle fibers are striated (darkly cross-striped). When the cardiac muscle fibers shorten or contract, they squeeze the chambers of the heart, thereby pushing out the blood they contain.

If we patiently cut all the way through the myocardium, the *endocardium* (**en**-doh-**KAR**-dee-um) is revealed (see Figure 11.3). The endocardium is the

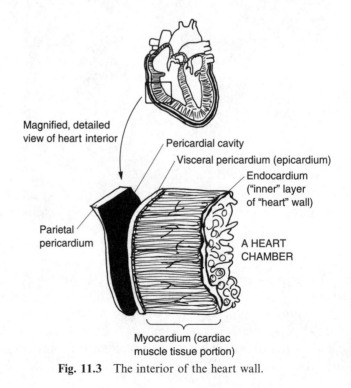

Fig. 11.3 The interior of the heart wall.

"inner" (*endo-*) layer of the heart wall. It forms the lining of the heart chambers. Like the epicardium (visceral pericardium), the endocardium is a *serous* (**SEER**-us) *membrane* – one that secretes a clear, "watery" (*ser*) fluid. This watery (serous) fluid helps moisten and lubricate the heart wall.

To summarize, the overall sequence of all heart layers (outside-to-in) is:

Parietal Pericardium $\longrightarrow$ **Visceral Pericardium (Epicardium)** $\longrightarrow$ **Myocardium**
$$\downarrow$$
Endocardium

The Blood Vessels and Their Walls

Just as the heart wall contains a number of distinct layers, so does the wall of each major blood vessel. Here, however, they are called *tunicas* (**TOO**-nih-kahs) – Latin for "coats or sheaths." This colorful name reflects the fact that *tunics* (**TOO**-niks) were worn like shirts or gowns by the Ancient Greeks and Romans. Figure 11.4 provides a cross-section (transverse section) through a capillary, artery, and vein. It shows the vessel openings, as well as the tunics or tunicas present as "coats" within the walls of the arteries and veins.

THE LUMEN LETS IN THE "LIGHT"

In the very center is the vessel *lumen* (**LOO**-men) or "light space." It is the hollow opening within the vessel which could (at least in theory) act as a passageway for light. It is via the lumen, of course, that the blood flows through the vessel.

ENDOTHELIAL CELLS LINE ALL BLOOD VESSELS

In all types of blood vessels, there is a single layer of *endothelial* (**en**-doh-**THEE**-lee-al) *cells*. The endothelial cells are flat, scale-like cells lying upon a basement membrane. The endothelial cells directly contact the vessel lumen. That's why their name begins with the prefix, *endo-* ("inner" or "innermost").

BLOOD CAPILLARIES HAVE NO TUNICAS

The capillaries are the tiniest blood vessels, and have the thinnest walls. Thus, they have no tunicas or coats. Each capillary is a very thin, "hairlike" vessel that directly supplies tissue cells with oxygen and other nutrients. It also picks up their metabolic waste products. The lumen of the capillary is often so

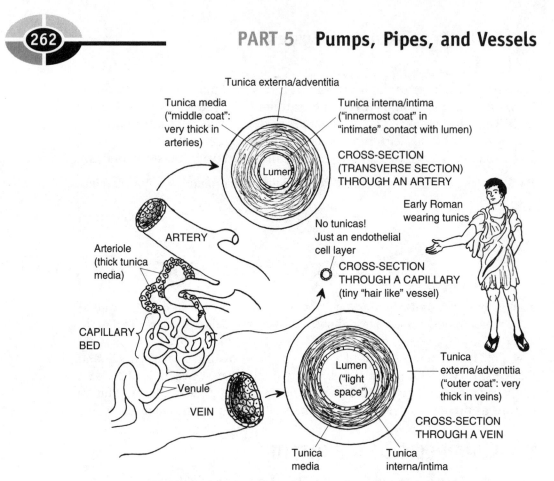

Fig. 11.4 A look at the major types of vessels and their walls.

narrow that only a single red blood cell can pass through it! And the capillary wall is only one cell layer thick – consisting of a single layer of endothelial cells upon a supporting basement membrane. Because they have such a thin wall, capillaries are the only vessels (other than venules) that allow diffusion of nutrients and waste products across them. Since they lack a rigid wall with tunicas, capillaries have a collapsible lumen and intermittent (on again, off again) blood flow. Hence at any given time, about 75% of the capillaries are collapsed and empty! [**Study suggestion:** Picture capillaries as tiny creeks running through the woods. In the spring, with lots of melting snow and rain, the creeks are full of water. But late in the summer, the creeks are usually dry, with no water flowing through them.]

THE THREE TYPES OF TUNICAS

Quite unlike capillaries, the walls of arteries and veins contain three tunicas:

1. The innermost tunica. The innermost tunica is called the *tunica interna* (in-**TER**-nah) or *tunica intima* (**IN**-tih-**mah**). It is the "innermost coat," which is named for its close, "intimate" contact with the vessel lumen. The tunica interna (tunica intima) includes the endothelial cells resting upon their basement membrane. In addition, there is also a thin layer of areolar (loose) connective tissue with stretchy elastic fibers, located around the endothelial cell lining. [**Study suggestion:** It would be a very good idea to go back and review the illustration of areolar or loose connective tissue, in Figure 5.8, so you can better appreciate its presence within the blood vessel wall.]

2. The middle tunica. The middle tunica is called just that – the *tunica media* (**ME**-dee-ah) or "middle coat." The tunica media is the thick coat of smooth muscle fibers within the wall of major vessels. Remember how we talked about smooth muscle as a type of visceral effector of the Autonomic Nervous System, near the beginning of this section? Well, the smooth muscle fibers in this layer get their "smooth" label because they are *non-striated* (**nahn-STREYE**-ate-ed) – "not (cross)-striped."

Nevertheless, smooth muscle fibers (just like their striated cousins, the cardiac and skeletal muscle fibers) have the ability to contract and shorten. This gives them an essential role in the movement of the blood vessel wall. In the case of the tunica media, the smooth muscle fibers form a circular ring or noose completely *around* the wall of the artery (see Figure 11.4).

Now, when you have seen cowboys get a rope or noose around their necks on TV shows, what eventually happens to them? "They get *hung!*" Baby Heinie quickly replies.

In the case of blood vessels, this tightening-of-a-noose event is called *vasoconstriction* (**vas**-oh-kahn-**STRIK**-shun), or the "process of vessel narrowing." As the tunica media smooth muscle fibers contract, then, they tighten the noose around the vessel lumen, thereby *vasoconstricting* (**vas**-oh-kahn-**STRIK**-ting) it. Conversely, if the smooth muscle fibers in the tunica media relax, then the noose opens a lot wider, and *vasodilation* (**vas**-oh-die-**LAY**-shun) occurs. Vasodilation exactly translates to mean the "process of vessel widening."

"Does this vasoconstriction stuff choke the blood vessel to death, like those bad cowboys who get hung in the movies, Professor?" No, Baby Heinie. What does happen, however, is an increase in the associated blood pressure (BP). Therefore, vasoconstriction tends to *raise* the BP, while *vasodilation* tends to *lower* it.

Another job done by the smooth muscle in the tunica media is regulating the amount of blood flowing into the vessels, downstream. After *vasoconstriction*, the volume of blood flowing downstream is *reduced*. Conversely, after *vasodilation*, the amount of blood flowing downstream is *increased*. [**Study**

suggestion: Using your new cardiovascular knowledge, explain why the two preceding sentences are true.]

Although the tunica media is mainly composed of smooth muscle tissue, there are also some collagen fibers and stretchy elastic fibers.

3. The outermost tunica. There is a third coat or layer in the wall of major blood vessels. It is called the *tunica externa* (eks-**TERN**-ah) or *tunica adventitia* (**ad**-ven-**TISH**-ee-ah). This coat is obviously the one forming the "outside" (*external* surface) of the vessel wall. Since the tunica externa is on the outer surface of the vessel, it is considered *adventitial* (**ad**-ven-**TISH**-al) i.e. "coming from the outside."

The tunica externa (adventitia) is mainly composed of *dense fibrous* (**FEYE**-brus) *connective tissue*, also called *collagenous* (kuh-**LAJ**-ih-nus) *connective* tissue. This type of connective tissue basically consists of a densely-packed collection of collagen fibers. Since collagen fibers are somewhat like ropes, they give the tunica externa (tunica adventitia) an essential role in anchoring the blood vessel wall to the surrounding body structures.

And in certain blood vessels (such as the veins), there is an added network of elastic fibers within the tunica externa. This helps make large veins very elastic and stretchy.

Thumbnail summary:

THE 3 TUNICAS = IN THE BLOOD VESSEL WALL	TUNICA *INTERNA* OR INTIMA (The *innermost* coat of endothelial cells lining vessel)	+	TUNICA *MEDIA* (The *middle* coat of smooth muscle fibers that *constricts* or *dilates* lumen)	+	TUNICA *EXTERNA* OR ADVENTITIA (The *outer* coat of collagen fibers that anchors and supports the vessel)

Tissue 2

General Structure–Function Characteristics of the Vessels

Now that we have briefly discussed *cardiac muscle* within the *heart* wall, and *smooth muscle* within the *blood vessel* wall, it is appropriate to examine the *general structure–function characteristics* of the blood vessels. And, after that, we will hook this vascular network back up to the heart.

THE ARTERIES ALWAYS COME *FIRST*, RIGHT *AFTER* THE HEART!

The arteries are literally "windpipes or air-keepers" (*arteri*). This strange name reflects the observations of the Ancient Greek and Roman anatomists. Take a quick glance back at Figure 11.1, and observe the sequence of vessels in the vascular network. The *arteries* always come *first*, carrying blood directly *away from* the heart. Since the contraction of the cardiac muscle in the myocardium creates the blood pressure, and arteries are right off the heart, the arteries are large-diameter vessels with a very high *blood pressure (BP)*. Now, the blood within the arteries moves very fast, and at a high blood pressure. Therefore, soon after death, the blood is pushed out of the arteries, and into the *venous* (**VEE**-nus) or "veiny" side of the vascular network. Hence, when the Ancient Anatomists cut open the lumen (light space) of the artery of a cadaver, it was empty and full of air! This is the strange story explaining how the concept of arteries as "windpipes or air-keepers" was born.

[**Study suggestion:** Visualize arteries as wide, raging rivers of red, foaming blood, violently thrashing and splashing with strong currents and great speed of flow.]

The artery wall

Take another peek at Figure 11.4, and concentrate upon some of the special features you find in the arterial wall, as compared to the vein wall. What especially sticks out is that the thickest tunica of the three – the tunica media – seems to be a lot *thicker* in the wall of the *artery*, compared to the *vein*. The *primary* component making up the tunica media comprises *smooth muscle fibers*, arranged in a *circle* around the vessel lumen. And when those muscle fibers *contract*, they cause vasoconstriction and an increase in the BP. When the smooth muscle fibers *relax*, they cause vasodilation and a decrease in BP. Therefore, since arteries have the thickest tunica media, they must make the most important contribution to regulation of the BP.

Different artery types

But we have to clarify this conclusion. It is the small-and-medium-sized arteries, usually called the *muscular or distributing arteries*, that are able to vasoconstrict and vasodilate enough to affect blood pressure significantly. They are called "muscular" because of the high proportion of smooth muscle in their walls, and they are called "distributing" because they are the arteries that carry (distribute) blood to all parts of the body, and its periphery. Most of the arteries, such as the *brachial* (**BRAY**-kee-al) *artery*, fall into this group. [**Study suggestion:** Using your knowledge gained from previous chapters on muscles and bones, in what *specific* part of the body would you say that the brachial artery lies? Why?]

"What about the really *big* arteries, Professor Joe?" Okay, Baby Heinie. Please take a minute and turn back to Figure 11.2, where we retracted some of the gross membranes covering the heart and its major vessels. Now, holding that page open in your book, do you see the *common pulmonary* (**PULL**-mah-**nair**-ee) *artery*, which is alternately called the *pulmonary trunk*? It is the most anterior major artery springing up from the base (flat top) of the heart. It is called the "common" pulmonary artery, or the pulmonary "trunk," because this very large artery serves as the "common trunk" or place of origin, for both the *right and left pulmonary* arteries.

They are called *pulmonary* arteries, because *pulmonary* literally "pertains to" (-*ary*) the "lungs" (*pulmon*). Therefore, the right and left pulmonary arteries carry the blood to the right and left lungs.

Now, if you look back at Figure 11.2, again, you should see another really huge artery. It is curving from right-to-left, just behind the common pulmonary artery (pulmonary trunk). It is called the *aortic* (ay-**OR**-tik) *arch*. This "arching" vessel is the first or proximal part of the *aorta* (ay-**OR**-tah), the main artery that literally "raises or lifts up" (*aort*) the blood out of the left side of the heart.

Both the common pulmonary artery (pulmonary trunk) and the aortic arch are classified as *elastic or conducting arteries*. Being the biggest arteries in the body, and coming directly off the heart, these vessels are subjected to a terrific amount of blood pressure! Their "elastic" name refers to the fact that their walls (especially the tunica externa) contain a lot more *elastic fibers*, than do other types of arteries. This allows the common pulmonary artery *(CPA)* and aortic arch to stretch and expand whenever the heart chambers push blood through them at high pressure, and then to *recoil* or snap back to *conduct* the blood farther out into the vascular system. But since their walls are so thick, the elastic (conducting) arteries can do very little vasoconstricting or vasodilating.

THE ARTERIOLES ARE THE "LITTLE" BRANCHES OF THE ARTERIES

Right after arteries in the vascular network, come the arterioles. The arterioles are literally just "little (branches of the) arteries" (*arteriol*). A look back at Figure 11.4, indeed, shows that the arterioles are a lot smaller in diameter than the arteries. The larger arterioles still have all three tunicas in their walls, with the tunica media (as in the wall of the arteries) being the thickest. The arterioles do become smaller and smaller in diameter as they travel ever farther from the heart.

The arterioles finally carry blood into the capillary network (which we have already discussed). Because of their position, then, they are important "choke points" for blood flow into the capillary network. Being thinner walled, the arterioles are able to vasoconstrict very powerfully, and reduce the size of their lumens dramatically. Thus, they can exert an important influence upon blood pressure and what is called *total peripheral resistance (TPR)* to blood flow. By *resistance (R)* to blood flow, we mean the ability to "take a stand back" or oppose the flow of blood through a vessel. When many arterioles vasoconstrict, they dramatically increase the total peripheral resistance (both in the body "periphery" and elsewhere). This great increase in TPR both significantly increases the blood pressure, as well as significantly reducing the flow of blood into the capillary networks.

[**Study suggestion:** Imagine that the arterioles are just smaller branches of the red, wild rivers – the arteries – but that they also have the ability to dramatically narrow their diameter at some times, and widen it at other times.]

THE VENULES ARE THE "LITTLE" BRANCHES LEADING *INTO* THE VEINS

Just as the arterioles bring blood *into* the capillary network, the *venules* carry blood *out* of the capillary network. Thus begins the long journey back towards the heart.

Venules are created when several capillaries merge together. The smallest venules, therefore, have pretty much the same structure as the capillaries – just a single layer of endothelial cells upon a basement membrane. The smallest venules (like the capillaries) allow diffusion, osmosis, and other types of particle movement across their walls. As the venules approach the veins, though, their walls get thicker. At their distal ends, the venules have a thin tunica media of smooth muscle in their wall.

In conclusion, we can say that the venules are built and function much *like* the *capillaries*, but that they *lead into* the veins. (Even though they are named "tiny veins," the venules do *not* simply represent a miniature version of a vein's anatomy.)

THE VEINS ALWAYS COME *LAST*, RIGHT *BEFORE* THE HEART!

The "little circle" of the human circulation takes its final curve through the veins. A vein, also called a *vena* (**VEE**-nah) in Latin, is a vessel that returns blood back *towards* the heart. For this reason, veins come *last* in sequence within the vascular network, right *before* the heart.

A patient glance back at Figure 11.4 shows that the veins have the same three tunicas as do the arteries. As you can see, though, veins have *wider lumens* compared to arteries, but *thinner walls*, overall. In particular, veins have a thinner tunica media than arteries. This means that they are much less able to vasoconstrict or vasodilate by contraction–relaxation of smooth muscle in their walls. (This also means that veins in a cadaver are much less stiff to the touch than are arteries, and that their lumens are readily collapsible.)

Although their total wall is thinner, veins have a thicker tunica externa (adventitia) than do arteries. This gives the veins a much higher proportion of collagen fibers. In addition, the large veins have an added network of elastic fibers.

Large veins, therefore, function mainly as *blood storage depots*. Having so many elastic fibers (as well as collagen fibers) in their tunica externa, they are able to expand and store considerable amounts of blood volume. (At any given time, more than 50% of the blood is being slowly carried through the veins.)

The two largest veins in the body can be seen back in Figure 11.2, which revealed the surface anatomy around the heart. The *superior vena cava* (**KAY**-vah) and the *inferior vena cava* – the "upper and lower cave or hollow veins" – are shown entering the right-hand side of the heart. The *superior* vena cava returns venous blood from the area *above* the heart, while the *inferior* vena cava returns blood from the entire body *below* the heart. Being so far from the pumping end of the heart, the two *vena cavae* (**KAY**-vigh) have an extremely low blood pressure, and their rate of blood flow is very slow. [**Study suggestion:** Picture each vena cava as a very wide, dark, slowly-moving river that stores a huge volume of fluid.]

In medium-sized veins of the arms and legs, there are *venous valves*. Each valve is a thin flap that opens in only one direction – towards the heart. Because veins are such low-pressure vessels, the venous valves help keep the blood from pooling (due to the effects of gravity) in the body extremities.

The Heart: A Double-Pump for Two Circulations

We have now come full circle – arteries to arterioles, arterioles to capillaries, capillaries to venules, venules to veins, and finally, veins back to the heart. It is time to look inside the heart, and study the chambers and other structures that either pump blood out into the arteries, or receive the blood slowly pouring in from the veins.

There are really two different circulations, each of them associated with only one side of the heart. The *pulmonary or right heart (RH) circulation* represents the circulation of blood to, through, and from the "lungs" (*pulmon*). The *systemic* (sis-**TEM**-ik) *or left heart (LH) circulation* represents the circulation of blood to, through, and from the organs of all the major body "systems," except for the lungs.

In summary, we have:

THE TOTAL BLOOD = **CIRCULATION**	**THE *PULMONARY* +** **(RH) CIRCULATION** *(Circulation of blood to, through, and from both "lungs")*	**THE *SYSTEMIC*** **(LH) CIRCULATION** *(Circulation of blood to, through, and from the organs of all body "systems" except for the lungs)*

Tissue 1

THE PULMONARY (RIGHT HEART) CIRCULATION

Observe from Figure 11.5 that the pulmonary or right heart circulation begins with the *right atrium* (**AY**-tree-um), a small "entrance room" (*atri*) "present" (*-um*) at the top of the heart. Externally, the chamber of the right atrium continues out into the *right auricle* (**AW**-rih-kl). The auricle is literally a "little ear" – a small, ear-like, hollow flap that beats like a wing on the outside of the atrium.

Actually, both the right atrium and its outer "little ear" (the right auricle) go into *systole* (**SIS**-toh-lee) or "contraction" together. The cardiac muscle tissue in their walls contracts, pressing upon the blood contained in their chambers.

From the right atrium (and its right auricle), the blood gets pushed down through the *tricuspid* (try-**KUS**-pid) *valve.* As its name indicates, the tricuspid valve is a one-way valve having "three" (*tri-*) leaf-like flaps with "points" (*cusps*). Thus, tricuspid valve is an anatomic name based upon the number

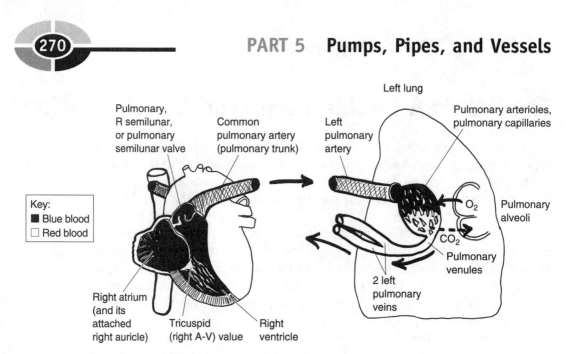

Fig. 11.5   An overview of the pulmonary (right heart) circulation.

and shape of the valve flaps. But the tricuspid valve has an alternate name –
the *right atrio-ventricular* (**ay**-tree-oh-ven-**TRIK**-yew-lar) or *right A-V valve*.

The alternate name for the tricuspid valve reflects its anatomic location. It
is the valve between the right "atrium" (*atrio-*) and the *right ventricle* (**VEN**-
trik-l). Each of the two ventricles (the right ventricle and the *left ventricle*) are
literally "little bellies"; i.e. little belly-like chambers near the bottom of the
heart.

The right ventricle is the major "lung pump" for the pulmonary circulation. After it fills with blood from the right atrium (through the open right A-
V valve), the right ventricle goes into systole. This contraction phase pumps
the blood up through the *pulmonary valve*, which is alternately called the *right
semilunar* (**sem**-ee-**LOO**-nar) *valve*. The word, semilunar, "pertains to" (*-ar*) a
"half" (*semi-*) "moon" (*lun*). Thus the pulmonary (right semilunar) valve
flaps are, indeed, shaped like "half-moons"! Anatomists often combine the
two different names for this valve (pulmonary valve or R semilunar valve)
into a third, combined name – the *pulmonary semilunar valve*.

"Do we *really* have to know *all three* of these valve names, Professor Joe?"
Well, Baby Heinie, it is certainly a good idea. In anatomy, I'm afraid, we just
have to get used to memorizing several names for body structures. That way,
we will recognize them if we hear or read about them from different sources
in the future.

As I was saying, the blood from the right ventricle pushes through the
pulmonary semilunar valve (pulmonary valve or R semilunar valve), and

goes up into the common pulmonary artery (pulmonary trunk). The blood then travels through either the right pulmonary artery, and into the right lung, or through the left pulmonary artery, and into the left lung.

Let's just look at what happens inside the left lung. The left pulmonary artery branches into a number of *pulmonary arterioles*. These, in turn, branch extensively into a series of *pulmonary capillaries*. You will note from Figure 11.5 that the blood is dark (actually a dark red or "blue"-red color) in the pulmonary circulation, all the way up through about the first half of the length of the pulmonary capillaries.

"How come it changes to a *lighter* color here, Professor?" Well, observe in Figure 11.5 that the pulmonary capillaries pass right alongside of the *pulmonary alveoli* (al-**VEE**-oh-**lie**). The pulmonary alveoli are "small hollow cavities" (*alveol*), actually extremely thin-walled tiny sacs in the lungs, that contain air. The fresh oxygen (O_2) in the air, diffuses out across the walls of the alveoli, and into the blood of the pulmonary capillaries. Picking up fresh O_2 changes the color of the blood from dark bluish-red, to a bright cherry-red, from the last half of the pulmonary capillaries, onward. Note, also, that carbon dioxide (CO_2) molecules diffuse out of the blood in the pulmonary capillaries, and into the alveoli.

At the ends of the pulmonary capillaries, they merge together to create *pulmonary venules*. The pulmonary venules, in turn, merge to create two *left pulmonary veins*. The same thing happens in the right lung, so there are also two *right pulmonary veins*. The four pulmonary veins (2 R pulmonary veins from the right lung, 2 L pulmonary veins from the left lung) finally end the pulmonary circulation.

To briefly summarize:

THE PULMONARY (RIGHT HEART) CIRCULATION:
Begins with the right atrium $\longrightarrow$ Goes to and includes the 4 pulmonary veins (and its right auricle)

Tissue 2

THE SYSTEMIC (LEFT HEART) CIRCULATION

We have just finished tracing the sequence of the pulmonary (right heart) circulation. The reason it was called the right heart circulation was because the right ventricle is the main pump that pushes blood up into both lungs.

Now let's examine its partner, the systemic or left heart circulation. As Figure 11.6 shows, this circulation begins with the *left atrium*, where the four pulmonary veins dump their blood. The *left auricle* is the outer "little ear" that assists the left atrium in pumping blood down through the *bicuspid*

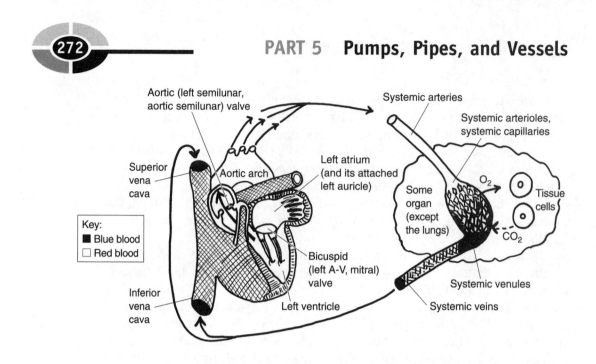

Fig. 11.6 An overview of the systemic (left heart) circulation.

(by-**KUS**-pid) *valve*. The bicuspid valve is a one-way valve having "two" (*bi-*) leaf-like flaps with "points" (*cusps*). Since the bicuspid valve is on the left side of the heart, it is also called the *left atrioventricular* or *left A-V valve*. In anatomy, of course, we have a multitude of names. Thus, the bicuspid or left atrioventricular (left A-V) valve is also known by a third name, the *mitral* (**MY**-tral) *valve*. This strange name comes from the fact that the valve closely resembles an inverted (upside-down) pointed hat of a bishop, which is called a *miter* (**MY**-ter). Hence, mitral literally "pertains to" (*-al*) a bishop's "miter" (*mitr*) (see Figure 11.7, A).

Blood is pushed through the bicuspid (left atrioventricular or mitral) valve, down into the *left ventricle*. The left ventricle is the main pump for the systemic circulation (including all the blood vessels in the entire body, except for those in the pulmonary circulation). The left ventricle powerfully contracts, sending blood up through the *aortic* (ay-**OR**-tik) *valve*. The aortic valve gets its name from the fact that it sits near the base of the aortic arch. Its alternate name is the *left semilunar valve*, since (like the right semilunar valve) its flaps are shaped much like half-moons. Thus the preferred name for this valve is the combined phrase, *aortic semilunar valve* (see Figure 11.7, B).

The blood goes up through the aortic semilunar valve, and into the aortic arch. Now, the aortic arch (and the lower portions of the aorta) send out numerous branches (smaller arteries), which either directly or indirectly

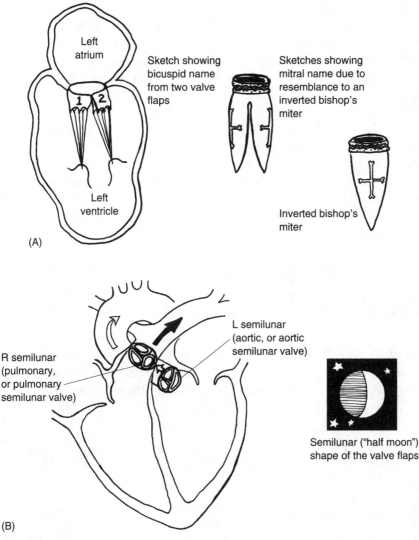

Fig. 11.7 Detailed anatomy and symbolism of the two types of heart valves. (A) A visual explanation for the alternate names of the left A-V valve. (B) A visual explanation for the names of the right and left semilunar valves.

supply all the organ *systems* of the body (except for the lungs). Consequently, we can provide a general name and call these the *systemic arteries*. Particular systemic arteries are usually named for the bones by which they pass, or the parts of the body through which they pass. Familiar examples would include the brachial artery (in the "arm"), and the *femoral artery* (in the "thigh," or along the "femur").

As these systemic arteries go into some organ (such as a particular bone, skeletal muscle, the liver, or the brain), they branch into many *systemic arterioles*. Like rivers that successively divide into smaller creeks, the systemic arterioles subdivide into numerous *systemic capillaries*.

The systemic capillaries, being extremely thin walled, dump their oxygen (O_2) molecules off to the tissue cells. As the O_2 molecules diffuse out of the blood, and into the tissue cells, there is a diffusion of CO_2 in the opposite direction. As a result, the blood in the second half of the systemic capillaries turns back to blue (actually dark, "bluish"-red) in color.

The systemic capillaries merge together to join with the *systemic venules*. The systemic venules (like all types of venules) run together to create the *systemic veins*. The systemic veins carry blood *away from* the body organs, and back towards the heart. The systemic circulation finally ends with the two largest systemic veins – the superior and inferior vena *cavae* (**KAY**-veye).

To briefly summarize:

THE SYSTEMIC (LEFT HEART) CIRCULATION:
Begins with the left atrium $\longrightarrow$ **Goes to and includes both the**
(and its left auricle)   **superior & inferior vena cavae**

Tissue 3

[**Study suggestion:** Go back and look through Figures 11.5 and 11.6. How many *total* one-way valves are present in the entire heart? How many of these are atrioventricular (A-V) valves? How many of these are semilunar valves?]

Quick Capsule Comments on the Veins and Arteries

Note from both Figures 11.5 and 11.6 that *arteries* always carry blood *away from* the heart. (In particular, they carry blood away from either the right or the left ventricle, which pumps the blood into them.)

Conversely, *veins* always return blood *back towards* the heart. They always go back toward either the right atrium or the left atrium – the two upper heart chambers which ultimately receive all of the *venous* (**VEE**-nus) blood.

"Professor, I've heard that *arteries* always contain *red* blood, and *veins* always contain *blue* blood! Is that *really* true?" To answer Baby Heinie's question, all we have to do is go back and carefully re-examine Figures 11.5 and 11.6. Now, in Figure 11.5, what color is the blood in the pulmonary *arteries*? "Gosh, it's dark blue!" And what color is the blood in the four pulmonary *veins*? "Goodness, it's bright cherry red!" Thus you can see that, for the *pulmonary* circulation, arteries contain blue blood, while veins contain red blood.

"Oh, no! Now I'm really mixed up, Prof Bony Joe! Are you saying that what I've heard all my life about *arteries* containing *red* blood, and *veins* containing *blue* blood, is *wrong*?"

Take a look back at Figure 11.6, Baby Heinie. Please observe that for the *systemic* circulation, what you have been told all your life about *arteries* containing *red* blood, and *veins* containing *blue* blood, is a true fact! All you have to remember about red-versus-blue blood in the future is to carefully distinguish whether you are talking about the blood in the *pulmonary* arteries and veins, or the blood in the *systemic* arteries and veins.

The Coronary Circulation: A "Special" Circulation of the Heart

We now arrive at the topic of *"special" circulations*: i.e. particular branches of the systemic circulation that supply blood to certain very "special" areas of the body.

THE CORONARY CIRCULATION "CROWNS" THE HEART

Ironically enough, one of the really "special" circulations of the body is the circulation of blood to the myocardium – the very cardiac muscle tissue within the wall of the heart, itself. This special circulation is technically called the *coronary* (**KOR**-uh-**nair**-ee) *circulation*. The coronary circulation is the circulation of blood to, through, and from the cardiac muscle within the wall of the heart.

P

Tissue 4

Figure 11.8 provides an abbreviated view of the coronary circulation. The term, coronary, literally "refers to" (-*ary*) a "crown" (*coron*). Our chief focus is upon the *right and left coronary arteries* (and their various branches). These vessels are named for their resemblance to a "crown" (*coron*) encircling the top of the heart, just below the auricles or "little ear"-like outer flaps. In short, the right and left coronary arteries look like a red-colored crown slipped down over the ears of a real prince! Their main job is to provide vital oxygen, glucose, and other nutrients to the extremely hardworking cardiac muscle tissue within the myocardium.

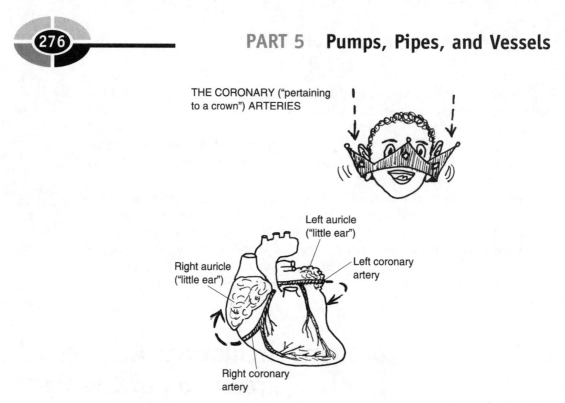

THE CORONARY ("pertaining to a crown") ARTERIES

Fig. 11.8 The coronary circulation: A little prince gets "crowned".

"Heart Attack" and Atherosclerosis

Just as long as the lumens of the coronary arteries are free and open, the cardiac muscle tissue is constantly provided with a flow of vital nutrients, as well as a circulatory pathway for excreting its metabolic waste products. The potential problem, however, is that the lumens of these absolutely critical arteries are only about as wide as a pencil lead!

This strange anatomic fact makes the coronary arteries dangerously vulnerable to the grave danger of *vessel occlusion* (uh-**KLEW**-zhun). Occlusion is literally a "closing up" (*occlus*). The main culprit that can cause such an occlusion or "closing up" of the coronary arteries is the lipid substance, cholesterol. (Review Chapter 4, if desired.)

Cholesterol is a white, fatty substance that tends to be laid down as *atheromas* (**ah**-ther-**OH**-mahs) or "fatty" (*ather*) "tumors" (*-omas*) whenever its blood concentration becomes too high. The extremely dangerous result is *atherosclerosis* (**ah**-ther-oh-sklair-**OH**-sis) – a "fatty hardening" (*-sclerosis*) of the arteries.

Organ 1

Figure 11.9 reveals the pathological anatomy of atherosclerosis. Excessive blood cholesterol (as well as several circulating blood fats) become deposited as atheromas upon the inner walls of arteries. Atherosclerosis mainly occurs within the lumens of the systemic arteries, which have a relatively high blood pressure. The high BP tends to push more cholesterol into fatty deposits on the endothelial cells lining these vessels.

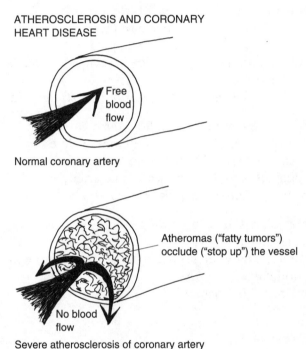

ATHEROSCLEROSIS AND CORONARY
HEART DISEASE

Free blood flow

Normal coronary artery

Atheromas ("fatty tumors")
occlude ("stop up") the vessel

No blood flow

Severe atherosclerosis of coronary artery

Fig. 11.9   Atherosclerosis and occlusion of the coronary arteries.

Atherosclerosis is especially harmful in cases of *coronary artery disease*, where atheromas can build up and suddenly occlude (close up) a coronary artery opening. As a result of this occlusion, the myocardium in the heart wall is suddenly choked off from oxygen, glucose, and other nutrients previously delivered by the coronary artery. The stricken person may experience crushing chest pain, a severe clinical symptom called *angina* (an-**JEYE**-nuh) *pectoris* (**PEK**-tor-is). A *coronary bypass operation* may have to be performed. In this operation, the surgeon bypasses the occluded parts of the coronary arteries by implanting small sections of veins obtained from other areas of the body. Hopefully, the operation works, and fresh blood is shunted past and around the blocked vessel areas to successfully feed the nutrient-starved cardiac muscle.

Quiz

Refer to the text in this chapter if necessary. A good score is at least 8 correct answers out of these 10 questions. The answers are listed in the back of this book.

1. The circulatory (cardiovascular) system consists of:
   (a) A "little circle" of digestive enzymes
   (b) A vascular network, but no pumping organs
   (c) Blood connective tissue, which is pumped by the heart through a vascular network
   (d) A series of vessels that simply run in a straight line

2. The correct sequence of blood flow through vessels:
   (a) Arterioles, veins, venules, capillaries, arteries
   (b) Veins, arteries, capillaries, venules, arterioles
   (c) Capillaries, venules, arteries, veins, arterioles
   (d) Arteries, arterioles, capillaries, venules, veins

3. The heart is anatomically located within the:
   (a) Mediastinum
   (b) Thoracic cavity, far to the right of the body midline
   (c) Abdominopelvic cavity
   (d) Upper vertebral canal

4. The outermost covering around the heart:
   (a) Myocardium
   (b) Parietal pericardium
   (c) Epicardium
   (d) Endocardium

5. The "light space" or opening within a blood vessel:
   (a) Lumen
   (b) Endothelial cells
   (c) Basement membrane
   (d) Central cavitation

6. Vessels that have no tunicas:
   (a) Veins
   (b) Capillaries
   (c) Arterioles
   (d) Arteries

7. The vessel "coat" mainly responsible for vasoconstriction:
   (a) Tunica intima
   (b) Tunica externa
   (c) Tunica media
   (d) Tunica adventitia

8. An elastic or conducting artery:
   (a) Brachial artery
   (b) Femoral artery
   (c) Tibial artery
   (d) Aorta

9. Always return blood back to the heart:
   (a) Coronary arteries
   (b) Auricles
   (c) Ventricles
   (d) Veins

10. The upper chambers of the heart:
    (a) Atria
    (b) A-V valves
    (c) Atherosclerosis
    (d) Semilunar valves

Body-Level Grids for Chapter 11

Several key body facts were tagged with numbered icons in the page margins of this chapter. Write a short summary of each of these key facts into a numbered cell or box within the appropriate *Body-Level Grid* that appears below.

Anatomy and *Biological Order* **Fact Grids for Chapter 11:**

TISSUE
Level

1	2

ORGAN
Level

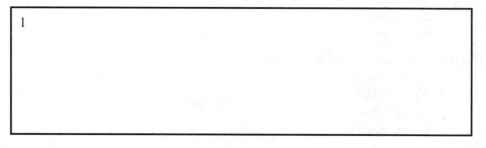

ORGAN SYSTEM
Level

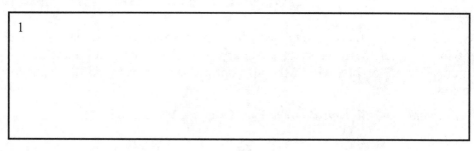

Physiology and *Biological Order* Fact Grids for Chapter 11:

TISSUE
Level

1	2
3	4

Physiology and *Biological Disorder* Fact Grids for Chapter 11:

ORGAN
Level

1

12

CHAPTER

The Anatomy of Blood and Lymph: Red Stuff and "Clear Spring Water"

In Chapter 11, we stated the basic identity of the circulatory (cardiovascular) system as follows:

A

CIRCULATORY (CARDIOVASCULAR) SYSTEM	=	THE HEART ORGAN	+	VASCULAR NETWORK	+	*BLOOD CONNECTIVE TISSUE*

Organ System 1   You will note in the above summary equation that only one item – the *blood connective tissue* – is specially emphasized. The reason is that we can put checkmarks of completion by the first two major components of the circulatory system (the heart organ and its attached vascular network). Only the *blood*, itself, was left out. Therefore, *here* is where we must begin!

The Blood: A Very "Special" Connective Tissue

Chapter 11 did define *blood connective tissue* as the material which is pumped in a "little circle" (*circul*) through the body's extensive network of blood vessels. And Chapter 5, you perhaps recollect, gave us a real *CNEMI* (**NEE**-mee) or "leg" up on the four basic types of body tissues. The first letter in *CNEMI*, of course, is C for *Connective Tissue*. It was pointed out that connective tissue either directly or indirectly connects body parts together.

Doesn't the blood, in fact, indirectly "connect" practically *all* of our body parts together? The reason, obviously, is that the blood is the *fluid* connective tissue that circulates throughout the human corpus, thereby linking body structures together (in a physiological sense rather than strictly anatomical one).

Blood is further classified as a "special" connective tissue because of the unique nature of the intercellular material between its cells – the *plasma* (**PLAZ**-muh). Plasma is the clear, watery, liquid "matter" (*plasm*) of both the blood and the *lymph* (**LIMF**) *connective tissues*. Therefore, after we examine the blood or "red stuff," we will be obliged to do the same for the lymph or "clear spring water."

To briefly capsulize, blood is a red, sticky connective tissue with a fluid intercellular matter or matrix (the plasma) that occupies about 4–6 liters of volume in the average-sized adult.

How Blood Looks in a Test Tube: The General Parts

"Professor, how can I tell anything about the blood, when it's moving through the blood vessels so darn fast?" Well, then we'll just have to "*prick* it 'n *spin* it"! Prick a finger with a sharp object and take a small sample of whole blood out of a blood vessel. Put the blood sample into a test tube, and spin the test tube in a *centrifuge* (**SEN**-trih-fyooj) machine. The centrifuge spins the blood around at a high rate of speed, so that the heavier portions of the blood fall down to the bottom of the test tube, and the lighter portions remain more towards the top (Figure 12.1).

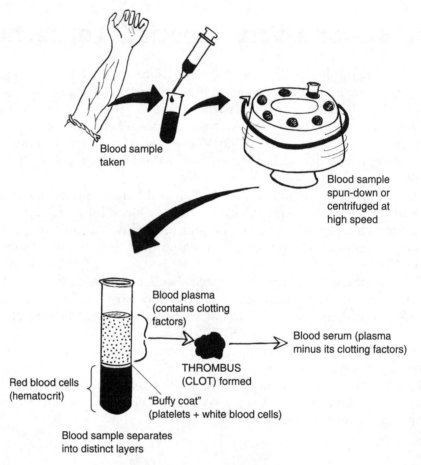

Fig. 12.1 General components of the blood.

STAGE #1 OF BLOOD ANALYSIS: THREE HORIZONTAL LAYERS

After centrifuging (spinning down of blood), we obtain three horizontal layers of general components of the whole blood tissue. These layers are the blood plasma, the clear, pale yellow, top layer making up 55% of blood volume; the *buffy coat*, the very thin, "buff"-colored (dull creamy yellow) middle layer occupying less than 1%; and the *red layer*, occupying the bottom 45% of the blood sample.

STAGE #2 OF BLOOD ANALYSIS: SEPARATING THE SERUM FROM THE PLASMA

If whole blood sits in a tube long enough, a *thrombus* (**THRAHM**-bus) or "clot" is soon formed. Around the thrombus (blood clot) is the *blood serum* (**SEER**-um). The word, serum, literally means "presence of whey" (*ser*) – the clear, watery by-product of cheesemaking. The blood serum, therefore, is the clear, watery portion of the blood plasma that still remains after a clot has been formed. In other words, the blood plasma minus its *clotting factors* results in the blood serum.

STAGE #3 OF BLOOD ANALYSIS: DISCOVERING OBJECTS CALLED THE "FORMED ELEMENTS"

The blood plasma (and serum) are basically just saline (saltwater) solutions containing a wide variety of dissolved solutes (atoms, ions, and molecules). Examining them under the compound light microscope will reveal few (if any) body structures that are larger or more complex than the ones occupying the Chemical Level of organization.

If the buffy coat and red layer are quickly viewed under a compound microscope, however, quite a different situation is observed! Here a great number of *formed elements* will be seen. By "formed elements," we mean blood cells and fragments of blood cells, which have their own plasma membrane, and hence, some distinctly visible "form" when viewed under the microscope.

QUICK SUMMARY OF INITIAL STAGES OF BLOOD ANALYSIS

Let us pool the above information about the beginning stages of blood analysis into several handy word equations:

A

Tissue 1

WHOLE BLOOD = **BLOOD PLASMA** + **THE BUFFY** + **THE RED**
(A CENTRIFUGED   *(Watery liquid*   **COAT**   **LAYER**
SAMPLE) (100%)   *"matter," 55%)*   **(Dull creamy**   **(45%)**
                                          yellow layer,
                                          less than 1%)

and

BLOOD PLASMA – CLOTTING FACTORS = THE BLOOD SERUM
                                    (Within a thrombus
                                    or "clot")

and

THE BUFFY COAT + THE RED LAYER = THE FORMED ELEMENTS
                                    (Blood cells and
                                    fragments of
                                    blood cells)

The Blood Plasma

Since no formed elements are seen within the blood plasma, when it is examined under a light microscope, we are essentially dealing with a large number of chemicals in a watery, fluid blood bath.

Blood plasma is basically a saline solution of 0.9% NaCl dissolved in water. Many other ions besides Na^+ and Cl^- are included, such as calcium (Ca^{++}) ions, *potassium (K^+) ions, hydrogen (H^+) ions*, and *bicarbonate* (buy-**KAR**-boh-nayt) or HCO_3^- *ions*. Blood plasma is a slightly *alkaline* (**AL**-kah-**lin**) or *basic solution*, having a *pH* of about 7.4. (The concept of pH and acid–base balance is discussed in the pages of *PHYSIOLOGY DEMYSTIFIED*.)

The plasma is full of dozens of different nutrients and minerals, such as glucose, iron, and cholesterol. Interestingly enough, a person's *blood profile* (received from an analysis by a medical laboratory) will usually report only *serum* values for the various chemical parameters in the bloodstream. We see items such as *serum glucose*, *serum iron*, and *serum cholesterol* levels, for example, rather than *plasma* concentrations for these same chemicals. Why? Apparently, the blood analysis is easier and cheaper if the blood plasma is allowed to clot, first, so that the various clotting factors (originally present within the plasma) are removed, leaving only the simpler blood serum to worry about.

There are numerous waste products of cell metabolism within the plasma, as well. Prominent among these are *urea* (you-**REE**-ah) and *uric* (**YOU**-rik) *acid*, which are both major nitrogen-containing waste products of protein catabolism (breakdown).

Blood plasma is the main carrier for most hormones (Chapter 10), so many hormone molecules will be found within any plasma sample. There are many enzymes in it, as well.

THE PLASMA PROTEINS

Although many blood-borne hormones (such as insulin) are proteins, and all enzymes (in the blood and elsewhere) are proteins, the majority of proteins found in the bloodstream are called the *plasma proteins*.

There are three broad types of plasma proteins: the *albumins* (al-**BYOO**-mins), *globulins* (**GLAHB**-you-lins), and the *clotting proteins*. All of them are important synthesis products of the liver.

The albumins

The word, albumin, comes from the Latin for "white of an egg." The albumins are thick, sticky, glue-like proteins found in raw egg white and in various other plant and animal tissues. The albumins are the most abundant of the plasma proteins, making up about 60% of the total. Because they are so abundant, their concentration can significantly affect the water concentration within the blood plasma. When the blood albumin concentration is unusually high, for instance, the water concentration within the bloodstream is unusually low. As a result, there is a net osmosis (diffusion) of water from the surrounding body tissues, into the blood plasma. Such significant movements of H_2O into and out of the bloodstream, have important influences upon both the blood volume and blood pressure.

The globulins

The word, globulin, exactly translates to mean a "little globe" (*globul*) "protein substance" (*-in*). This rather rounded, globe-shaped group makes up about 1/3 (36%) of all the plasma proteins. There are several subtypes of globulins, usually named with some of the letters of the Greek alphabet. The *alpha* and *beta globulins*, for instance, are transport proteins that carry lipids, certain ions, and fat-soluble (dissolvable) vitamins. Another main subtype is the *gamma* (**GAM**-ah) *globulins*. These serve as *antibody molecules* that attack and destroy foreign invaders as a critical part of the body's *immune* (ih-**MYOON**) or self-defense system.

The clotting proteins

The last main subtype of plasma proteins comprises the clotting proteins, which make up about 4% of the total. The two most important clotting proteins are called *fibrinogen* (feye-**BRIN**-oh-jen) and *prothrombin* (proh-**THRAM**-bin). Fibrinogen and prothrombin are present in the blood plasma all the time.

They undergo a sequence of reactions to help produce blood clotting, whenever there is some damage or injury to the blood vessel wall.

In summation:

A

$$\begin{array}{llll} \textbf{THE PLASMA} = & \textbf{ALBUMINS} & + \textbf{GLOBULINS} + & \textbf{CLOTTING} \\ \textbf{PROTEINS} & \textbf{\textit{(Osmosis, blood}} & \textbf{\textit{(Transport,}} & \textbf{PROTEINS} \\ & \textbf{\textit{volume, BP)}} & \textbf{\textit{immunity)}} & \textbf{\textit{(Blood clotting)}} \end{array}$$

Molecule 1

The Buffy Coat: "Yes, It's *Buffy*! But Is It Also *Fluffy*?"

The second or middle layer of blood components in a centrifuged tube is the buffy coat. When carefully examined through a light microscope, the explanation for the dull, yellow-whitish color that gives the buffy coat its name becomes quite obvious. The two major formed elements within the buffy coat are called the *leukocytes* (**LEW**-koh-**sights**) and the *platelets* (**PLAY**-teh-**lets**) or *thrombocytes* (**THRAHM**-buh-**sights**).

THE LEUKOCYTES OR "WHITE CELLS"

The leukocytes are literally the "white" (*leuk*) blood "cells" (*cytes*). These cells typically have a large, purplish-staining nucleus, but they are named for the clear, whitish appearance of their cytoplasm (see Figure 12.2). (Their presence in such large numbers within the buffy coat, obviously, helps give this centrifuged layer its creamy yellowish-white appearance.) The leukocytes play significant roles in protecting the body from various foreign invaders.

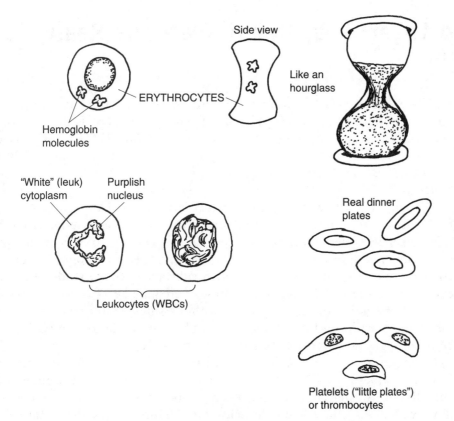

Fig. 12.2   The major types of formed elements in the blood.

THE PLATELETS – NOT JUST A BUNCH OF DIRTY DISHES!

The second type of formed element within the buffy coat is the platelet or thrombocyte. The platelet name comes from their shape – "little plate"-like fragments of disintegrated bone marrow cells. The thrombocyte name derives from their function – "clotting" (*thromb*) "cells." About 1/3 the size of an RBC (red blood cell), the platelets (thrombocytes) are scattered here and there in small groups, throughout the plasma. These purplish-colored, plate-like cell fragments have very sticky surfaces. Thus, whenever a vessel ruptures and *hemorrhages* (**HEM**-eh-**rij**-es) or "bursts forth" blood, the platelets soon collect around the open hole and help create a thrombus (clot). This physiological behavior in forming clots explains the platelets' alternate name of thrombocytes.

The Red Layer: Yes, Our Hemato*crit* Really Is *Crit*ical!

Finally, we are ready to consider the third and widest layer that appears at the very bottom of a centrifuged tube of blood. We called it the red layer, because it consists of thousands of *erythrocytes* (air-**RITH**-roh-**sights**) or "red" (*erythr*) blood "cells" (*cytes*). Erythrocytes (red blood cells or RBCs) are *anucleate* (**ay-NEW**-klee-aht); that is, they are "without" (*a-*) any "nucleus."

They are also quite special in that they are shaped like *biconcave* (buy-**KAHN**-cave) *discs*, being "caved-in" on "both" (*bi-*) sides. Viewed from the side, this makes them look like red hourglasses! The red color is mainly due to the presence of *hemoglobin* (**HEE**-moh-**glohb**-in). Hemoglobin is a reddish-colored, "globe"-shaped (*glob*), "protein substance" (*-in*) found within the cytoplasm of the red "blood" (*hem*) cells. There may be as many as 250–280 million hemoglobin molecules present within a single erythrocyte! The main job of these millions of hemoglobin molecules is carrying oxygen (O_2) molecules through the bloodstream, and to the tissue cells.

An important term involving the erythrocytes is *hematocrit* (he-**MAT**-oh-krit). Hematocrit literally means the "separation" (*crit*) of a test tube filled with "blood" (*hemat*) into its three main components or layers, by spinning it down with a centrifuge. Being the densest of the three layers, the erythrocytes "separate" (*crit*) from the other two layers, settling into a *packed volume of red blood cells, only*.

In other words:

A

Cell 1

THE HEMATOCRIT = (OF A CENTRIFUGED BLOOD SAMPLE)	THE VOLUME OF RBCs "SEPARATED" (CRIT) OUT AND PACKED AT BOTTOM OF TEST TUBE IS STATED AS A CERTAIN % OF THE TOTAL BLOOD VOLUME

Note from the above equation that the hematocrit of a given person is expressed as some percentage of their total volume of blood. The normal range for hematocrit is 43–49% of the total blood volume in men; and slightly less, 37–43% of the total blood volume, in women. The average or set-point value for hematocrit is usually given as about 45% (or nearly half) of the total volume of blood. Since erythrocytes are so critical for carrying oxygen through the bloodstream, it should be obvious that, as stated at the beginning of this section, "Yes, our hemato*crit* really is *critical*!" [**Study**

suggestion: Would you rather have a hematocrit value near the *Upper Normal Limit* of its range for you as a male or female, or one near its *Lower Normal Limit* for you? Explain.]

Lymph Formation: "Don't Cry over Spilled *Blood*! − It's the *Lymph*! Now, Is That *Clear*?"

Near the very beginning of this chapter, we defined *plasma* as the clear, watery, liquid "matter" (*plasm*) of both the blood connective tissue and the *lymph connective tissue*. Hence, we need to discuss the basic anatomy and characteristics of the lymph, just as we have done for the blood.

The word, *lymph*, exactly translates to mean "clear spring water." Lymph is a usually clear, transparent, colorless fluid (like clear spring water). So it is pretty much equivalent to the blood plasma in most places of the body, except that its content of proteins is much lower. The lymph, then, is the clear, watery, liquid intercellular material of the *lymphatic circulation*.

THE LYMPHATIC SYSTEM: A SHADOW CIRCULATION OF THE BLOOD

The last chapter or two have focused upon the circulation of the *blood*. However, the blood circulation is being *followed*! It is being followed by something like a *shadow*! It is being followed by the *lymphatic* (lim-**FAT**-ik) *circulation*, a major part of the *lymphatic system*.

As overviewed in Chapter 3, the lymphatic system is an organ system that consists of a collection of *lymphatic organs* and *lymphatic vessels* running between them. In a practical sense, the lymphatic system can be thought of as a shadow circulation of the blood circulation. The main reason is that the *lymphatic capillaries* run side-by-side (like a shadow) along the tiny blood capillaries (Figure 12.3).

The lymphatic capillaries are special in that they are dead-ended. Since they are closed at the far end, their fluid contents – the lymph – flows in one direction only, towards the heart. The lymph (which looks like clear spring water) is actually a *filtrate* (**FILL**-trait) – filtration product – of the blood, the plasma in the nearby blood capillaries. The blood pressure (BP) pushes outward, thereby filtering water, NaCl, and various foreign objects (such as dirt, bacteria, or cancer cells) out of the blood and into the lymphatic capillaries.

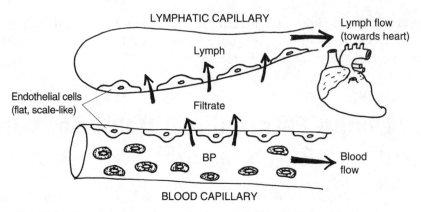

Fig. 12.3 Lymph is formed as a shadow of the blood. BP = blood pressure.

Since the lymph usually contains no erythrocytes (RBCs), it looks clear, rather than red-colored. But eventually, the lymphatic vessels drain their cleansed lymph into several major blood veins that flow into the top of the heart.

The Lymphatic System Linked to Immunity

Immunity (ih-**MYOO**-nih-tee) is literally a "condition of" (*-ity*) "not serving" (*immun*) disease. More directly, immunity is a condition of freedom or protection from disease. This state of immunity is closely associated with the functioning of the lymphatic system. Hence, Chapter 3 introduced the lymphatic system as actually being a combined organ system, called the *lymphatic-immune* (ih-**MYOON**) *system*.

The *immune system* can be considered one aspect or facet of the lymphatic system. The immune portion of the lymphatic is involved in such activities as the production of chemical antibodies (remember the gamma globulins?).

Therefore, we can modify the summary equation that appeared back in Chapter 3, and obtain this new expression for the combined lymphatic-immune system:

A

| **LYMPHATIC-IMMUNE =** **SYSTEM** | **LYMPHATIC** **SYSTEM** *(Lymphatic vessels & organs containing lymph)* | **+** | **IMMUNE** **SYSTEM** *(Antibodies and other protectors from disease)* |

Anatomy to the Rescue:
The Reticuloendothelial System

The lymphatic system is sometimes given the alternate name of *reticulo-endothelial* (reh-**TIK**-you-loh-**en**-doh-**THEE**-lee-al) *system*. The extremely long word, reticuloendothelial, is often simply abbreviated as *R-E*. And the reticuloendothelial system simply becomes named the *R-E system*.

The reticuloendothelial or R-E system is a "little network" (*reticul*) of lymphatic vessels that are lined by "endothelial" cells. So, it is a name based upon anatomy. (You may remember the endothelial cells from Chapter 11 as the flat, scale-like cells that also line the lumens of our blood vessels.) [**Study suggestion:** Take another quick look back at Figure 12.3, and observe how the endothelial cells line both the blood capillaries and the lymphatic capillaries.] Because these endothelial cells are so flat, lymph (and its contained dirt, bacteria, or cancer cells) are readily filtered through or between them.

Since the lymphatic (reticuloendothelial) system receives dirt, bacteria, debris, viruses, and cancer cells that have been filtered out of the blood-stream, it serves a critical role in immunity (body defense) by cleaning up the blood, and then returning the cleansed fluid back to the blood. Remember this general saying: "From the blood the lymph is formed, and back to the blood the lymph doth return." The first lymph filtered from the bloodstream is dirty (in the sense that it often carries contaminants or foreign invaders), while the final lymph is clean (in the sense that the contaminants and foreign invaders have been removed). Therefore, in the process of filtering and cleaning the material that is leaked out of the blood capillaries, the lymphatic (reticuloendothelial) system provides immunity.

We can summarize the important inter-relationships by two alternate equations:

**LYMPHATIC SYSTEM = RETICULOENDOTHELIAL (R-E) SYSTEM
= IMMUNE SYSTEM**

or

LYMPHATIC-IMMUNE SYSTEM = R-E SYSTEM

A

Organ System 3

Major Lymphatic Organs and Lymphatic Tissue

The vessels of the R-E system or network travel to and from some major lymphatic organs. These are sketched in Figure 12.4.

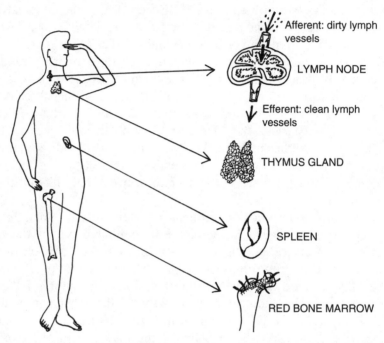

Fig. 12.4 Some major lymphatic organs.

THE LYMPH NODES

The most widespread lymphatic organs are the *lymph nodes*. The lymph nodes are a group of small, bean-like organs scattered in clusters in various parts of the body. *Afferent* (**AF**-fer-**ent**) *lymphatic vessels* "carry" (*fer*) dirty lymph "towards" (*af-*) the lymph nodes. *Efferent* (**EE**-fer-ent) *lymphatic vessels* carry clean lymph "away from" (*ef-*) them.

THE THYMUS GLAND

A most unusual lymphatic organ is the *thymus* (**THIGH**-mus) *gland*. The word, thymus, comes from the Ancient Greek for "warty outgrowth," reflecting the bumpy appearance of this endocrine gland. The thymus is a thin, flat gland lying just deep to the sternum or "breastplate." This gland consists of two bumpy-looking lobes.

The thymus gland is most prominent in young humans and other mammals. (In calves and lambs, it is called the "throat sweetbread," because it is often eaten as sweet-tasting meat.) *Thym* also means "sweet," like the leaves of the *thyme* plant. And the "bread" is *roasted meat*. Hence, Aristotle dissected calves to obtain the thymus or "sweetbread" (Figure 12.5). The thymus gland reaches its maximum size at puberty, then progressively decreases in size. In most adults, the thymus is completely gone, having been replaced by connective tissue.

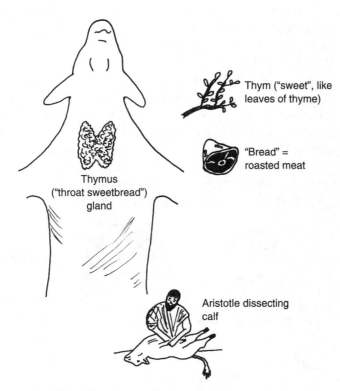

Thym ("sweet", like leaves of thyme)

"Bread" = roasted meat

Thymus ("throat sweetbread") gland

Aristotle dissecting calf

Fig. 12.5   The thymus gland as the "throat sweetbread."

RED BONE MARROW

The red bone marrow found within spongy bone (Chapter 7) is a third important type of lymphatic organ.

SPLEEN

In humans, the *spleen* is a dark red organ attached to the left side of the stomach. It looks somewhat like a thick, crescent-shaped roll (croissant), and it rather feels like one, being soft and spongy to the touch.

THE TONSILS: "LITTLE ALMONDS" IN THE BACK OF OUR THROAT

In addition to full-blown lymphatic organs, there are smaller masses of lymphatic tissue scattered here and there around the body. Prominent among these are the *tonsils* (**TAHN**-sils). The tonsils are literally "little almonds" – oval, somewhat almond-shaped clusters of lymphatic tissue – lying in the back of the throat (Figure 12.6).

There are five tonsils. The *pharyngeal* (fah-**RIN**-jee-al) *tonsil* is the single uppermost mass, located in the portion of the *pharynx* (**FAIR**-inks) or "throat" just behind the nose. The pharyngeal tonsil is also called the *adenoids* (**AD**-uh-**noyds**), because it is rather big and "gland" (*aden*) "like" (*-oid*).

The two *palatine* (**PAL**-ah-**tyn**) *tonsils*, as their name indicates, are a pair of tonsils lying on either side of the throat, just below the *palate* (**PAL**-aht) or "roof of the mouth." Finally, there is a pair of *lingual* (**LING**-gwal) *tonsils*, attached way back at the base of the "tongue" (*lingu*).

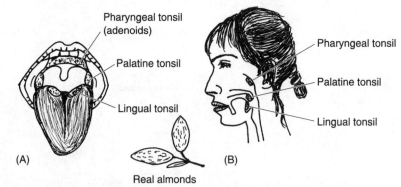

Fig. 12.6   The tonsils: "Little almonds" tucked away in our throats.

A

LYMPHATIC TISSUE SUMMARY

Organ System 4   Here is a concluding listing of all the major types of lymphatic organs and tissues:

LYMPHATIC   = LYMPH + THYMUS + RED BONE + SPLEEN + TONSILS
ORGANS/TISSUE   NODES   GLAND   MARROW

Tonsillitis

Since they are composed of lymphatic tissue, the five tonsils play minor roles in body defense. Sometimes, however, the lymphatic tissue of the tonsils becomes overwhelmed by a huge number of invading bacteria or viruses. In such cases, *tonsillitis* (**tahn**-sihl-**EYE**-tis) may result. Tonsillitis is "an inflammation and swelling of" (-*itis*) the tonsils. This inflammation may be accompanied by a dangerously high fever. The operation of *tonsillectomy* (**tahn**-sihl-**EK**-toh-mee) or "removal of" (-*ectomy*) the tonsils is then frequently performed. [**Study suggestion:** Using your growing knowledge, write a single term that literally means, "inflammation of the adenoids." Why do you think that a person afflicted with this condition might have trouble breathing?]

A

Tissue 1

Quiz

Refer to the text in this chapter if necessary. A good score is at least 8 correct answers out of these 10 questions. The answers are listed in the back of this book.

1. Blood is a type of connective tissue in that it:
    (a) Includes slender fibers that strap or connect body parts together
    (b) Circulates throughout the body, hence "functionally" connects the parts
    (c) Has a microscopic anatomy much like bone connective tissue
    (d) Removes waste products via the kidneys

2. The fluid intercellular substance of blood:
    (a) Erythrocytes
    (b) Buffy coat
    (c) Leukocytes
    (d) Plasma

3. The so-called buffy coat in centrifuged blood consists of:
    (a) Leukocytes and platelets
    (b) Spleen cells and red bone marrow
    (c) A frothy foam of lipids and circulating blood fats

    (d) Erythrocytes and thrombocytes

4. What is meant, exactly, by the "hematocrit"?
    (a) The critical mass of lymphatic tissue needed for immunity
    (b) The volume of RBCs that "separate" from the rest of the blood when a sample is centrifuged
    (c) The percentage of critical nutrients in the blood, as opposed to the lymph
    (d) A large number of circulating white blood cells different from the others

5. The blood serum equals:
    (a) Blood plasma minus clotting factors
    (b) The RBCs plus the WBCs
    (c) Blood plasma plus many clotting factors
    (d) The dark bluish part of the blood connective tissue

6. The most abundant group of plasma proteins:
    (a) Globulins
    (b) Hemoglobin
    (c) Fibrinogens
    (d) Albumins

7. The lymphatic circulation can be called a shadow circulation because:
    (a) The contained lymph and other fluid is dark (like a shadow)
    (b) When malfunctioning, it makes us walk with a "limpf"
    (c) The lymphatic capillaries tend to run parallel to the blood capillaries
    (d) It is an imaginary circulation, having no actual physical reality

8. The lymphatic system is closely linked to immunity, in that it:
    (a) Circulates antibodies and specific cells involved in body defense
    (b) Never becomes inflamed or attacked by foreign organisms
    (c) Appears in the body only when immune activities are in progress
    (d) Is the main cause for most endocrine diseases

9. The reticuloendothelial (R-E) system involves a:
    (a) Little network of lymphatic vessels lined by flat endothelial cells
    (b) Spherical group of plasma proteins arranged into striations
    (c) Large group of erythrocytes clumped into a "little mesh"
    (d) Tube of lymphatic organs critical for digestion of carbohydrates

10. A crescent-shaped lymphatic organ attached to the left side of the stomach:
    (a) Croissant muffin
    (b) Thymus
    (c) Spleen
    (d) Thyroid

Body-Level Grids for Chapter 12

Several key body facts were tagged with numbered icons in the page margins of this chapter. Write a short summary of each of these key facts into a numbered cell or box within the appropriate *Body-Level Grid* that appears below.

Anatomy and *Biological Order* **Fact Grids for Chapter 12:**

A

MOLECULE
Level

1

CELL
Level

1

TISSUE
Level

1

ORGAN SYSTEM
Level

1	2
3	4

Anatomy and *Biological Disorder* Fact Grids for Chapter 12:

TISSUE
Level

1

CHAPTER

It's All About The Respiratory "Trees": Their Branches Sway in the Airy "Breeze"

The last couple of chapters have dwelled upon the blood – its *pumps* (the chambers of the heart), its *pipes* (the various types of blood vessels), and its *plasma* (the fluid and formed elements within blood connective tissue). There was even talk about a vascular network – a complex, tree-like arrangement of numerous branching blood vessels. Blood, then, is considered a fluid substance: one which can be compressed, pushed through pipes, and carried for long distances.

The Respiratory System: To "Air" Is Only Human!

In a quite similar manner, we can talk about the air. The air, like the blood, is a fluid substance, and it, too, can be compressed, pushed through pipes, and carried for long distances. This fluid air is what we breathe into and out of our *respiratory* (**RES**-pih-rah-**toh**-ree) *system*. The very word, respiratory, does in fact, translate to mean "pertaining to" (*-ory*) "breathing" (*spirat*) "again and again" (*re-*).

The respiratory system, therefore, is the organ system devoted to breathing again and again. And as a result of this breathing process, fresh oxygen (O_2) molecules are delivered to the bloodstream, while accumulated carbon dioxide (CO_2) molecules are carried away from the bloodstream.

THE TWO MAJOR DIVISIONS OF THE RESPIRATORY SYSTEM

Anatomically speaking, the respiratory system can be conveniently studied by breaking it into two major divisions – an *Upper Respiratory Tract* at the superior end of the breathing pathway, and a *Lower Respiratory Tract* at the inferior end:

THE RESPIRATORY =	THE UPPER +	THE LOWER
(*"Breathing*	RESPIRATORY	RESPIRATORY
again-and-again")	TRACT	TRACT
SYSTEM		(*The Respiratory "Tree"*)

A

Organ System 1

THE CONCEPT OF A RESPIRATORY "TREE"

Like the vascular tree or network of blood vessels, the Lower Respiratory Tract consists of hollow tubes that branch repeatedly, much like a tree. Hence, the Lower Respiratory Tract is sometimes called the *Respiratory Tree*. It is a branching network of numerous air tubes, having various thicknesses and diameters.

It will be our main task in this chapter to help you clearly understand the sequence of major respiratory system structures that air passes through, as it winds its way through both the Upper Respiratory Tract, as well as the Lower Respiratory Tract or Respiratory Tree.

The Upper Respiratory Tract – Holes and Tubes Above the "Tree"

The extensively branching network of the respiratory tree doesn't begin immediately. Rather, the first thing we encounter within the *respiratory system* is the *Upper Respiratory Tract* (see Figure 13.1). The Upper Respiratory Tract can be defined as the superior portion of the respiratory system, which is literally involved in "breathing" (*spir*) "again and again" (*re-*). Speaking informally, the Upper Respiratory Tract consists of the holes and tubes above the branching portion of the respiratory sytem – the respiratory tree.

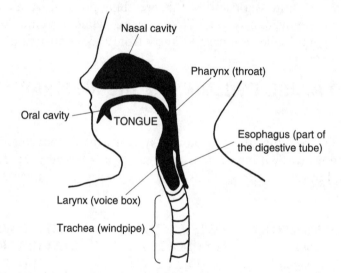

Fig. 13.1 The upper respiratory tract: Our airway down through the trachea (sagittal view).

To capsulize:

UPPER RESPIRATORY TRACT = Mostly Unbranched Holes and Tubes Lying above the Respiratory Tree, and Involved in Breathing

THE NOSE AND ITS NATURAL AIR FILTERS

The first places where we usually "breathe again and again" are through the *nasal* (**NAY**-sal) *cavity* and the *oral* (**OR**-al) *cavity*. The nasal cavity is the large hollow space within the "nose" (*nas*), while the oral cavity is situated just behind the lips and teeth of the "mouth" (*or*).

A sticky *mucous* (**MYEW**-kus) *membrane* lines the nasal cavity. It secretes *nasal mucus* (**MYEW**-kus) or "nose slime" (*muc*). Working with the nose hairs, the nasal mucus captures dirt and debris sucked into the nasal cavity during *inspiration* (**in**-spir-**AY**-shun). Inspiration is literally the "process of" (*-tion*) "breathing" (*spir*) "into" (*in-*) the body. The nasal hairs and mucus act like natural air filters during inspiration.

A rich vascular network lies just deep to the mucous membrane. The hot blood coursing through the vessels in this network serves to warm the air as it is being inhaled through the nostrils. The nasal mucus provides additional help by moistening the inspired air as it moves through the nasal cavity.

After the nasal and oral cavities, inspired air goes into the *pharynx* (**FAIR**-inks) or "throat" (*pharyng*).

THE LARYNX AND ITS LID

Situated at the inferior end of the pharynx, one finds the *larynx* (**LAIR**-inks) (see Figures 13.1 and 13.2.) The larynx or "voice box" (*laryng*) is a box-shaped collection of cartilage plates held together by dense fibrous connective tissue. Like the bow (tapered front end) of a ship, the *laryngeal* (lah-**RIN**-jee-al) *prominence* is a projection of cartilage sticking out from the front of the voice box. The laryngeal prominence is considerably larger in males than in females, due to the physiological influence of their higher levels of the hormone *testosterone* (tes-**TAHS**-ter-**ohn**). [**Study suggestion:** If you happen to be a *male*, put down this book and go watch yourself swallow in a mirror! If you happen to be a *female*, put down this book and go *watch* a friendly *male* swallow! That big, rounded bump you see traveling up-and-down in the front of the male neck is nicknamed the *Adam's apple*, but it is really just the laryngeal prominence of the voice box, covered over with flesh and skin!]

The larynx is nicknamed the voice box, of course, because it produces the sounds of the "voice" (*voc*). Stretched across the hollow interior of the larynx are the two *vocal* (**VOH**-kal) *cords* or "voice strings." The vocal cords are twin straps of highly elastic connective tissue. Being so elastic and stretchy, they strongly vibrate with the passage of air through the larynx. These vibrations create the *vocal* ("pertaining to voice") *sounds*.

Between the two vocal cords is a tapered, "tongue" (*glott*)-shaped opening called the *glottis* (**GLAHT**-is). Closely related to the glottis is the *epiglottis* (**EH**-pih-**glaht**-is). The epiglottis is a highly flexible flap of cartilage located "upon" (*epi-*) the glottis. The epiglottis thus serves as a flexible lid over the top of the larynx or voice box. When a person swallows, the food or liquid normally passes down the pharynx (throat) and pushes the epiglottis shut.

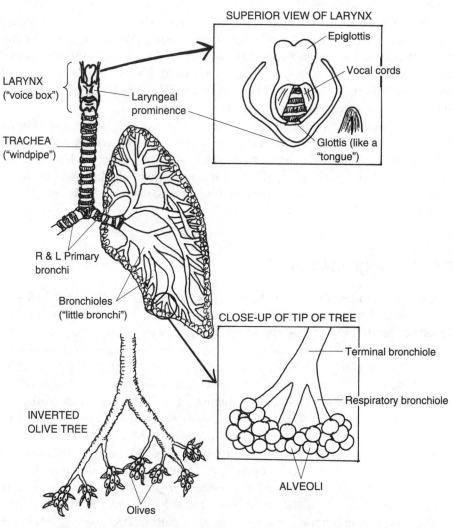

Fig. 13.2   The larynx down to the respiratory tree (anterior view).

This shutting mechanism normally prevents *pathologic* (*path*-oh-**LAJ**-ik) *respiratory aspiration* (as-pih-**RAY**-shun). This is the technical term for the "disease-causing," potentially dangerous, "breathing in" (*aspir*) of food, liquid, or other foreign objects into the larynx and airways.

Just below the larynx sits the *trachea* (**TRAY**-kee-ah) or main "windpipe." The trachea conducts air both into, and out of, the *right and left lungs*. The lumen (opening or light space) of the trachea is stiff and noncollapsible, due to the presence of partial rings or horseshoes of cartilage within its walls. (We are calling these horseshoes because they are partial rings of cartilage that encircle the trachea, but are open at the back like real horseshoes.)

The trachea is lined by a mucous membrane containing *ciliated* (**SIL-ee-ayt**-ed) epithelial cells. The *cilia* (**SIL-ee-ah**) are tiny, hairlike projections on the free (unattached) borders of the epithelial cells lining the trachea. Like the "eyelashes"(*cilia*) that help protect our eyes, the cilia help protect the lungs and deeper airways from inhaled dust, dirt, or debris. The mucous film inside the trachea is a sticky surface that captures and temporarily holds inhaled dirt and debris. The cilia constantly beat in an upward manner, thereby progressively moving the dirt-laden mucus in a superior direction, towards the pharynx. Once in the pharynx, the dirty mucus can either be swallowed, or spat out.

Organ 1

Summary. Let us now bring together the main components of the Upper Respiratory Tract. Thus:

UPPER RESPIRATORY TRACT	= NASAL CAVITY	+ ORAL CAVITY	+ PHARYNX (*"Throat"*)	+ LARYNX (*"Voice box"*)	+ TRACHEA (*Main "windpipe"*)

The Lower Respiratory Tract: Our Branching Respiratory "Tree"

The trachea travels along the body midline (midsagittal plane) as a single wide-diameter tube. At its caudal end, however, the trachea splits into two main branches. These are called the *right and left primary bronchi* (**BRAHNG-keye**). Each *primary bronchus* (**BRAHNG-kus**) looks very much like its mother-tube, the trachea. The primary bronchi, like the trachea, for instance, have wide, noncollapsible lumens held open by horseshoes of cartilage in their walls. They are also lined by a wet, sticky mucous membrane with a helpful ciliated epithelium. Hence, they are called *primary* ("first-order") *bronchi* ("windpipes").

THE LUNGS AND THEIR SUBDIVISIONS

Each primary bronchus enters the medial border of a lung through a *pulmonary* (**PUL-moh-neh-ree**) *hilum* (**HIGH-lum**). This is a "trifle" (*hil*) little slit in the "lung" (*pulmon*), through which major air tubes, blood vessels, and nerves enter and leave the organ.

The lungs, themselves, are two cone-shaped, spongy, pinkish-colored organs located within the thoracic cavity, and flanking the heart on either side. The right lung contains three *lobes* (major sections), while the left lung has only two lobes. These lobes are separated from each other by thin sheets of connective tissue. [**Study suggestion:** Recall that, in most people, about 2/3 of the heart lies to the left of the body midline. Does this give you any ideas about possible explanations for the differences in the number of lobes between the lungs?]

Basically, every structure in the respiratory system lying beyond the right and left primary bronchi is a part of the lungs and their lobes. Each lobe of the lung, in turn, is subdivided by many *lobules* (**LAHB**-yools) or "little lobes," separated by sheets of connective tissue. Each lobule is shaped like a hexagon, and is the smallest subdivision of the lung that is still visible to the naked eye.

THE BRONCHIAL (RESPIRATORY) TREE

A glance back at Figure 13.2 should refresh your mind as to what is meant by the respiratory "tree." Technically speaking, the respiratory tree is called the *bronchial* (**BRAHNG**-kee-al) *tree* – "pertaining to the bronchi." The reason is that most of the branches of this treelike network of air tubes are really bronchi of various sizes.

We are speaking, of course, about an inverted (upside-down) tree. Now, don't the larynx and trachea look somewhat like an inverted tree trunk with rough bark? Nevertheless, anatomists generally regard the respiratory tree as the bronchial tree. So, it formally begins with where the tree trunk (trachea) begins to fork into its two main branches – the right and left primary bronchi.

After entering a lung, each primary bronchus extensively branches into a series of progressively smaller bronchi. The primary bronchus is eventually followed by almost two dozen smaller bronchi. Rather than try to list all of them, we will just collectively call all of them the *smaller bronchi*.

Appearance of the bronchioles

As the bronchi become smaller and smaller, they progressively lose the cartilage horseshoes within their walls. Cilia and mucus thin out, then disappear. A layer of *circular smooth muscle tissue* (named for the fact that it forms a circle or noose around the tube lumen) gets thicker and thicker within the airway wall.

Finally, after the smallest branches of the bronchi narrow down to less than 1 mm (millimeter) in diameter, these tiny branches are called the *bronchioles* (**BRAHNG**-kee-ohls). The bronchioles are literally the "little" (*-ole*) "bronchi." But, as we have already pointed out, their wall structure differs considerably from that of the larger bronchi.

The bronchioles, like their bigger cousins, the bronchi, just keep branching and branching into smaller and smaller sets of bronchioles. When the tube lumen becomes less than 1/2 (0.5) of a millimeter in diameter, the bronchioles are called the *terminal bronchioles*.

Approaching the alveoli

Each terminal bronchiole, however, doesn't quite "terminate" or end the branching process. A close-up view of the tip of the respiratory tree (shown in the lower box back in Figure 13.2) reveals just a little more.

Several *respiratory bronchioles* branch off each terminal bronchiole. These respiratory bronchioles are, at last, the smallest of *all* the bronchioles! "Why are these tiniest of all bronchioles called the *respiratory* bronchioles, Professor Joe?"

The reason is that these final bronchioles are the type closest to the *respiratory membrane*, also called the *air–blood barrier*. And for inhaled air to get to this special respiratory membrane, it has to travel through the respiratory bronchiole, and, through the *very last branch* of the respiratory tree!

It is called the *alveolar* (al-**VEE**-oh-lar) *duct*, and as the name indicates, it is a tiny *duct* that travels deep into an *alveolar sac* (see Figure 13.3). [**Study suggestion:** Visualize the alveolar sac as a bunch of grapes. The alveolar duct is the small stem leading into the bunch of grapes.]

An alveolar sac is just a cluster of neighboring *pulmonary alveoli* (al-**VEE**-oh-lie). Each *pulmonary alveolus* (al-**VEE**-oh-lus) is literally a "little cavity" (*alveoli*) within the "lung" (*pulmon*). Doesn't each alveolus look somewhat like a hollow grape (see Figure 13.3) attached to a hollow stem (an alveolar duct)?

There are about 300 million alveoli contained within our two lungs. The alveoli are very special in their anatomy, in that their wall only consists of a single layer of endothelial cells resting upon a basement membrane. "Oh, isn't that the same thing we said about the wall of the blood capillaries, in an earlier chapter?" That is very correct! This thin barrier allows oxygen (O_2) and carbon dioxide (CO_2) molecules to diffuse freely across the alveolar wall, into and out of the bloodstream.

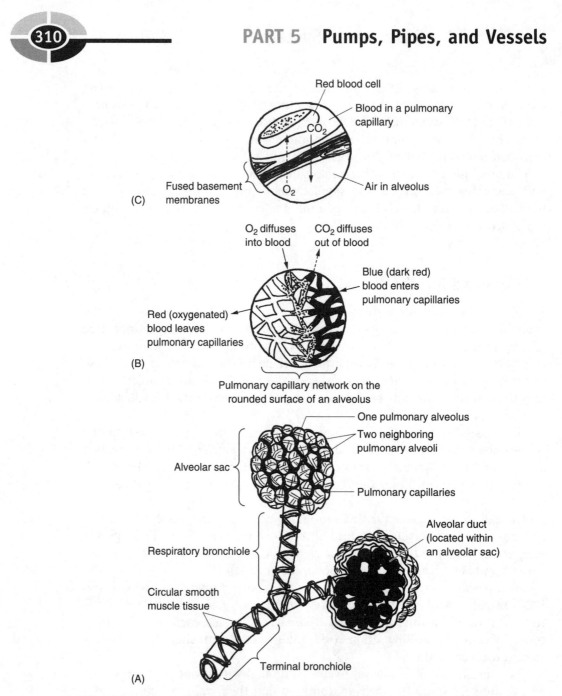

Fig. 13.3 Extreme close-up of the alveoli. (A) The final airways leading into the alveolar sacs. (B) A pulmonary capillary network changes its color from blue to red blood (due to addition of O_2 from the alveolus, and exit of CO_2 from the bloodstream). (C) Most highly magnified view: The actual respiratory membrane (air–blood barrier between an alveolus and pulmonary capillary).

SUMMARY OF THE RESPIRATORY TREE

We have finally gone through all the major parts of the respiratory (bronchial) tree, also called the Lower Respiratory Tract. A peek back at Figure 13.2 shows our overall impression of this lung-embedded tree as being somewhat similar to an inverted olive tree. In this case, of course, the alveoli are compared to clusters of olives (instead of grapes).

To briefly update, we have:

LOWER	=	***THE RESPIRATORY (BRONCHIAL) TREE:***
RESPIRATORY		**PRIMARY BRONCHI + SMALLER**
TRACT		**BRONCHI + BRONCHIOLES + ALVEOLAR**
		DUCTS + PULMONARY ALVEOLI

A

Organ 2

The Respiratory Membrane: The *Only* Site of Pulmonary Respiration

We have already mentioned the idea of a respiratory membrane, or the air–blood barrier. The respiratory membrane is the extremely thin barrier between the air within the cavity of each alveolus, and the blood within the lumen of an adjacent pulmonary capillary. The wall of the alveolus, plus the wall of the neighboring pulmonary capillary, together make up the respiratory membrane. The walls of the alveolus and its pulmonary capillary are so close, in fact, that the basement membranes of each one are fused together! (Check back with the extreme close-up, in Figure 13.3, C.)

OCCURRENCE OF TRUE PULMONARY RESPIRATION

The respiratory membrane (fused alveolar wall–pulmonary capillary wall) is the only place in the entire respiratory system where true *pulmonary respiration* occurs!

"Wait a minute, Prof! I thought the whole Lower Respiratory Tract was called the *respiratory* tree!" That's right, Baby Heinie. "Okay, so doesn't the word, respiratory, mean 'pertaining to respiration'?" That's right, Baby Heinie. "And didn't we say, earlier, that respiration literally means 'the process of breathing again and again'?" That's right, Baby Heinie. "So, doesn't this mean that the *whole* respiratory tree is involved in respiration – *not* just the alveoli?" That's *wrong*, Baby Heinie! I think the main reason you are confused, here, is due to a problem with using the literal or exact translation of the word, respiration.

Pulmonary ventilation versus respiration

In practical use, respiration is the process of *gas exchange* between two or more body compartments (such as the interior of the alveoli and the interior of the pulmonary capillaries). In pulmonary respiration, specifically, O_2 molecules diffuse from the air within each alveolus, and into the pulmonary capillary. Conversely, CO_2 molecules diffuse in the opposite direction – from the blood in the pulmonary capillary, into the air of an alveolus. There is consequently an *exchange of gases* (O_2 and CO_2), which are moving in *opposite directions* across the fused alveolar–capillary wall.

"I see, now, Prof! But why can't pulmonary respiration (diffusion of O_2 in one direction, CO_2 in the other direction) occur across the walls of the *rest* of the so-called respiratory tree – like across the walls of the bronchi and bronchioles, into and out of the bloodstream?" [**Study suggestion:** If you have been following this dialogue between Professor Joe and Baby Heinie carefully, you should be able to answer the above question. Try it! It's just common sense! Then check your thinking as you continue reading.]

The walls of all the other parts of the respiratory tree are just too thick to allow respiration to occur, because respiration occurs by the chance, random diffusion of oxygen and carbon dioxide molecules across the walls. And we know that diffusion is only effective over very short distances.

"Okay, I understand. But what process *is* really going on in the rest of the so-called respiratory system, if it's *not* respiration?" The answer is *pulmonary ventilation* (**ven**-tih-**LAY**-shun). Ventilation is literally "a process of" (*-tion*) "fanning or blowing air" (*ventil*). Pulmonary ventilation, therefore, is the process of sucking air into the lungs, and blowing of air out of the lungs. The main job of the entire respiratory system except for the pulmonary alveoli, then, is pulmonary *ventilation* – *not* pulmonary *respiration*!

SUMMARY OF RESPIRATION VERSUS VENTILATION

PULMONARY RESPIRATION = Gas Exchange (*by Simple Diffusion*)
between Air in Alveoli and Blood
in Pulmonary Capillaries

Organ 1   **while**

PULMONARY VENTILATION = *The Sucking of Air into the Lungs,*
                          And the Blowing of Air out of the Lungs

The Bronchiole and Its Muscular Wall

Our primary slant in the last few pages has been upon the pulmonary alveoli and the pulmonary capillaries, and their critical role in creating the respiratory membrane which allows true respiration to occur. The other airways were treated as perhaps a little less important.

BRONCHOCONSTRICTION VERSUS BRONCHODILATION: A CHOKEHOLD ON THE AIRWAYS

But the bronchioles, in particular, deserve our further attention. Back in Figure 13.3, the large amount of circular smooth muscle in the walls of the bronchioles is clearly illustrated. Recall also that the walls of the bronchioles lack the stiffening horseshoes of cartilage found in their more proximal neighbors, the bronchi.

Therefore, the bronchioles are very capable of changing the diameters of their lumens. A big change in bronchiole diameter will, of course, dramatically affect the pulmonary ventilation of the alveoli – the critical place where actual gas exchange with the blood occurs.

When a person is very stressed, or is vigorously exercising, the sympathetic portion of the Autonomic Nervous System (Chapter 9) is dominant in its influence upon visceral effectors (smooth muscle, cardiac muscle, and glands). The sympathetic nerves inhibit the circular smooth muscle tissue in the walls of the bronchioles. This causes the muscle to relax, and the bronchioles to widen. This physiological event is technically called *bronchodilation* (**brahng**-koh-dih-**LAY**-shun) – the "process of" (*-tion*) "bronchial tube" (*bronch*) "widening" (*dilat*). When the bronchioles widen, their resistance to air flow decreases; hence, there is a greater flow of air into and out of the alveoli. [**Study suggestion:** Why is this event of bronchodilation especially important when sympathetic nerve activity dominates within the body?]

Conversely, when a person is resting, relaxing, or digesting food, the parasympathetic portion of the ANS tends to dominate. Its effect upon the bronchiolar smooth muscle is one of stimulation. This stimulation causes the circular smooth muscle to contract, thereby narrowing the bronchiole lumen. This effect is called *bronchoconstriction* (**brahng**-koh-kahn-**STRIK**-shun) or "the process of narrowing of the bronchial tubes." Obviously, then, bronchoconstriction tends to reduce the flow of air into and out of the alveoli, since there is an increased resistance offered by the narrowed bronchiole tube lumen. In effect, we can think of bronchoconstriction as a noose tightening around the neck of an alveolus – its bronchiole. But, during resting conditions, a mild bronchoconstriction really isn't such a big chokehold, is it? [**Study suggestion:** Pretend that *you* are Professor Joe, and try to explain to Baby Heinie, using your own words, why mild bronchoconstriction when the parasympathetic nerves are active really does no harm.]

SUMMARY EQUATIONS

P

STRESS/EXERCISE:

$$SYMPATHETIC \longrightarrow INHIBITION\ OF \longrightarrow BRONCHO\text{-} \longrightarrow INCREASED$$
$$NERVES \qquad\qquad BRONCHIOLE \qquad DILATION \qquad AIR\ FLOW$$
$$ACTIVE \qquad\qquad SMOOTH \qquad\qquad\qquad\qquad TO\ ALVEOLI$$
$$MUSCLE$$

Organ 2   **versus**

REST/DIGESTION:

$$PARASYMPATHETIC \longrightarrow EXCITATION\ OF \longrightarrow \quad BRONCHO\text{-} \longrightarrow DECREASED$$
$$NERVES \qquad\qquad BRONCHIOLE \qquad CONSTRICTION \qquad AIR\ FLOW$$
$$ACTIVE \qquad\qquad SMOOTH \qquad\qquad\qquad\qquad TO\ ALVEOLI$$
$$MUSCLE$$

Bronchospasm and Severe Bronchitis: A Real "Chokehold" on Ventilation!

P

Organ 1

Although a mild-to-moderate degree of bronchoconstriction is a completely normal event, *bronchospasm* (**BRAHNG**-koh-spazm) is quite another matter! A *spasm* in general is any "convulsion" – a powerful, sudden sequence of involuntary muscle contraction, then relaxation. Bronchospasm, then, is a powerful and sudden contraction of the circular smooth muscle in the bronchiole wall. If strong and sustained enough, bronchospasm can produce a deadly chokehold of complete bronchoconstriction of important airways.

Bronchospasms often accompany both *asthma* attacks and *bronchitis* (brahng-**KEYE**-tis) or "inflammation of the bronchial tubes." Asthma is literally a "panting." In asthma, the affected person's immune system provides a strong allergic reaction to some inhaled substance – such as dust or pollen. Mild-to-moderate bronchoconstriction is a normal protective reflex response when a person accidentally inhales some poisonous or irritating substance (such as second-hand cigarette smoke). The bronchoconstriction is protective, in that it allows less of the inhaled substance to flow down into the alveoli.

But in asthma, the airways are too severely bronchoconstricted, often with powerful bronchospasms. As a result, the person "pants" (*asthma*), breathing rapidly and shallow in an effort to inhale enough air.

In bronchitis, the lining of the bronchi and bronchioles are all swollen and inflamed (perhaps from a bacterial or viral infection). This severe irritation often causes a *hypersecretion* (excessive secretion) of mucus, which also tends to block or occlude the extremely narrowed airways.

In such severe cases, where there is a real "chokehold" on pulmonary ventilation, *bronchodilator* (**brahng**-koh-**DIE**-lay-ter) drugs, such as epinephrine (adrenaline), may be given. The epinephrine (you may remember) is associated naturally with both the adrenal medulla (Chapter 10) and the sympathetic nerves. When given by injection, or inhaled as a mist, additional epinephrine from outside the body can achieve basically the same "bronchiole-widening" effects.

Quiz

Refer to the text in this chapter if necessary. A good score is at least 8 correct answers out of these 10 questions. The answers are listed in the back of this book.

1. The word, respiration, exactly translates from Latin to mean:
    (a) Curdling
    (b) The process of breathing again-and-again
    (c) Once-over, lightly
    (d) The process of exchanging gases between two or more body compartments

2. After the nasal and oral cavities, inspired air travels directly into the:
    (a) Alveoli
    (b) Larynx
    (c) Stomach
    (d) Pharynx

3. The technical name for the "Adam's apple" in males:
    (a) Vocal sounds
    (b) Pharynx
    (c) Glottis
    (d) Laryngeal prominence

4. The epiglottis serves as the:
    (a) Membrane containing both vocal cords
    (b) Flexible lid over the larynx
    (c) Supporting cartilage for the walls of the voice box
    (d) "Producer of the voice"

5. Each ____ ____ enters the medial border of a lung:
    (a) Pulmonary venule
    (b) Respiratory bronchiole
    (c) Primary bronchus
    (d) Alveolar duct

6. The respiratory structure having 3 lobes:
    (a) Trachea
    (b) L lung
    (c) Pulmonary alveolus
    (d) R lung

7. Contain a layer of circular smooth muscle that allows them to dramatically constrict:
    (a) Bronchioles
    (b) Primary bronchi
    (c) Mucous membranes
    (d) Pulmonary hilums

8. The tiniest of all the bronchioles:
    (a) Respiratory
    (b) Alveolar
    (c) Terminal
    (d) Proximal

9. Pulmonary ventilation involves:
    (a) Gas exchange by diffusion
    (b) Osmosis through selectively permeable lung membranes
    (c) Active transport of O_2 into tissue cells
    (d) The sucking of air into the lungs, and the blowing of air out

10. The fused walls of the pulmonary alveoli and pulmonary capillaries:
   (a) Respiratory membrane
   (b) Blood–brain barrier
   (c) Nasal septum
   (d) Pulmonary lobules

Body-Level Grids for Chapter 13

Several key body facts were tagged with numbered icons in the page margins of
this chapter. Write a short summary of each of these key facts into a numbered
cell or box within the appropriate *Body-Level Grid* that appears below.

Anatomy and **Biological Order** Fact Grids for Chapter 13:

A

ORGAN
Level

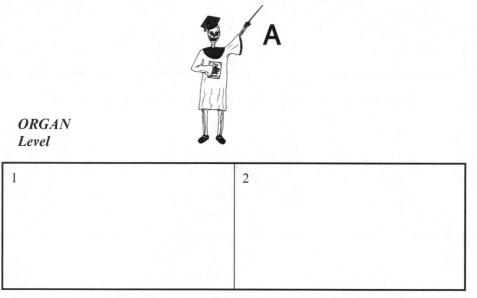

1	2

ORGAN SYSTEM
Level

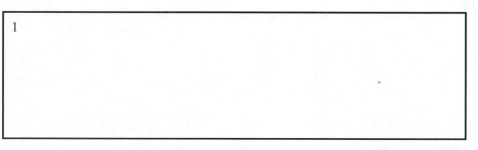

1

Physiology and *Biological Order* Fact Grids for Chapter 13:

ORGAN
Level

1	2

Physiology and *Biological Disorder* Fact Grid for Chapter 13:

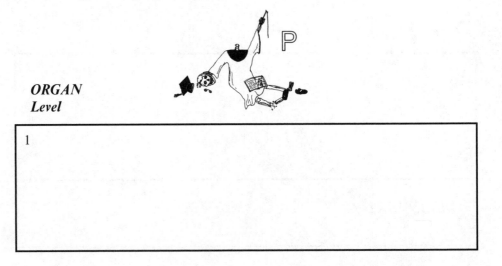

ORGAN
Level

1

Test: Part 5

DO NOT REFER TO THE TEXT WHEN TAKING THIS TEST. A good score is at least 18 (out of 25 questions) correct. Answers are in the back of the book. It's best to have a friend check your score the first time, so you won't memorize the answers if you want to take the test again.

1. The circulatory (cardiovascular) system equals the:
   (a) Heart plus vascular network plus blood connective tissue
   (b) Blood plus lymph plus renal fluid
   (c) Heart, lungs, blood, and blood vessels
   (d) Stomach, groin, liver, and veins
   (e) R-E system and the heart–lymph connection

2. The tiniest type of blood vessels:
   (a) Arteries
   (b) Venules
   (c) Capillaries
   (d) Veins
   (e) Arterioles

3. Cardiac muscle tissue lies within the ____ in the heart wall:
   - (a) Parietal pericardium
   - (b) Myocardium
   - (c) Endocardium
   - (d) A-V valves
   - (e) Visceral pericardium

4. Flat, scale-like cells lining the interior of all blood vessel walls:
   - (a) Glandular
   - (b) Mucous
   - (c) Endothelial
   - (d) Epithelial
   - (e) Fibroblasts

5. The outer coat of a vessel consisting largely of dense fibrous connective tissue:
   - (a) Pericardium
   - (b) Tunica externa or adventitia
   - (c) Parietal pleura
   - (d) Tunica media
   - (e) Ligamentum

6. Always carry blood away from the heart:
   - (a) Venules
   - (b) Arteries
   - (c) Veins
   - (d) Venous sinuses
   - (e) Lymphatic ducts

7. The common pulmonary artery and aortic arch belong to this general classification:
   - (a) Elastic or conducting arteries
   - (b) Peripheral veins
   - (c) Muscular or distributing arteries
   - (d) Capillary network
   - (e) Lymph-carrying tubes

8. Always carry blue blood:
   - (a) All veins
   - (b) Pulmonary arteries
   - (c) Systemic arteries
   - (d) All arteries
   - (e) Pulmonary veins

9. The two largest veins in the body:
   (a) Hepatic and abdominopelvic
   (b) Superior and inferior vena cavae
   (c) Femoral and sciatic
   (d) Jugular and common carotid
   (e) Brachial and axillary

10. The circulation of blood to, through, and from both lungs:
    (a) Systemic
    (b) Hepatic
    (c) Pulmonary
    (d) General
    (e) Coronary

11. The plasma is special because it:
    (a) Like the lymph, is a fluid intercellular substance
    (b) Contains unusually large quantities of sodium
    (c) Has no cells suspended within it
    (d) Transports oxygen, but not CO_2
    (e) Has a bright cherry-red color

12. Always settle at the bottom of a centrifuged tube of blood:
    (a) Platelets
    (b) Erythrocytes
    (c) Leukocytes
    (d) Plasma cells
    (e) Antibodies

13. The group of plasma proteins involved in immune reactions:
    (a) Globulins
    (b) Fibrins
    (c) Albumins
    (d) Fibrinogens
    (e) Prothrombins

14. The leukocytes get their name from the fact that:
    (a) They blanche to a pale yellow color when mixed with oxygen
    (b) The entire cell stays white-colored in albino individuals
    (c) Cytoplasm continually "leuks" out across the plasma membrane
    (d) Their nuclei are purplish, but their cytoplasm is clear
    (e) A lot of germs are destroyed by their actions

15. Anucleate, biconcave discs:
    (a) Thrombocytes
    (b) RBCs
    (c) T-lymphocytes
    (d) B-lymphocytes
    (e) Plasma cells

16. A hematocrit of 45% would suggest that a patient:
    (a) Had about an average percent of erythrocytes within the blood
    (b) Was suffering a severe and possibly fatal hemorrhage
    (c) Really had an unusually high ability to fight invading bacteria
    (d) Circulated blood with a reduced ability to clot
    (e) Showed a strong tendency towards cell lysis

17. A clear filtrate of the blood plasma:
    (a) Fibrinogen
    (b) Lymph
    (c) Soda pop
    (d) Serum
    (e) Agglutinin

18. Achieving a state of immunity suggests that:
    (a) The probability of suffering morbidity is greatly reduced
    (b) A person is approaching a condition of mortality
    (c) The lymphatic system is opposing the R-E network
    (d) Blood is circulating in the wrong direction
    (e) Homeostasis has been permanently maintained

19. Afferent lymphatic vessels:
    (a) Carry clean lymph towards the lymph nodes
    (b) Circle around-and-around the spleen, but never go inside of it
    (c) Merely agitate the lymph up-and-down, rather than transport it
    (d) Carry dirty lymph towards the lymph nodes
    (e) Don't really carry any lymph at all!

20. "Little almonds" of lymphatic tissue located in the back of the throat:
    (a) Pancreatic islets
    (b) Nut-so's
    (c) Tonsils
    (d) Pineals
    (e) Mammaries

21. Commonly known as the respiratory "tree":
    (a) The Upper Respiratory Tract

(b) Terminal bronchioles, only
(c) Laryngeal prominence
(d) The Lower Respiratory Tract
(e) The trachea

22. The process of breathing air into the lungs:
    (a) Respiration
    (b) Expiration
    (c) "Panting"
    (d) Inspiration
    (e) Perspiration

23. The tongue-shaped opening between the vocal cords:
    (a) Glottis
    (b) Adam's apple
    (c) Facquat's tomato
    (d) Epiglottis
    (e) Eve's orange

24. "Eyelash"-like projections that sweep away dirt-filled mucus:
    (a) Rough ERs
    (b) Flagellas
    (c) Squamous-shaped endothelial cells
    (d) Cilia
    (e) Hilums

25. The approximate total number of pulmonary alveoli present within both lungs:
    (a) 6,500
    (b) 150,000
    (c) 1,290,456
    (d) 10,000,000
    (e) 300,000,000

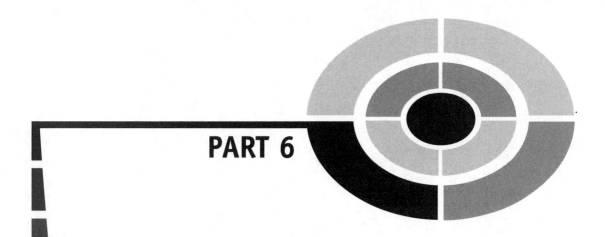

"Down-Under" Organ Systems, or the Body Land "Below Our Belt"

CHAPTER

14

The Digestive Tube: A Rabbit-Hole Outside the Body!

Part 5 talked about pumps, pipes, and vessels for blood and air. Now in Part 6, we are going to visit Australia. "Why Australia, Professor Joe?" – Because it's The Land "Down-Under"! As far as the human body is concerned, we have two "Down-Under" Organ Systems that are largely located in the Body Land "Below Our Belt." Specifically, these are the *digestive system* (Chapter 14) and the *genitourinary* (**JEN**-ih-toh-**ur**-ih-**nair**-ee) *system* (Chapter 15).

"Baby Heinie Just Swallowed a Marble! – *Where* Did It Go?"

A Anatomically speaking, the digestive system (also commonly known as the *digestive tube*) is a tube that extends from the mouth (oral cavity) all the way down to the *anus* (**AY**-nus). The anus is the small, muscular "ring" (*an*) through which one defecates.

Organ System 1

Digestive System = Tube from the Oral Cavity down to Anus

As you look at Figure 14.1, it would be useful to speculate about our mischievous little pal, Baby Heinie. Back a few years ago, in earlier childhood,

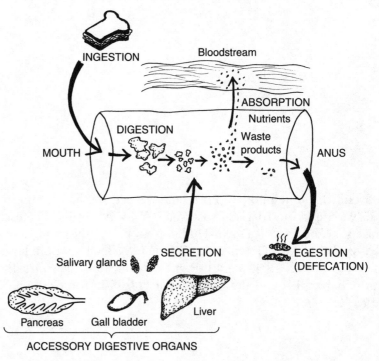

Fig. 14.1 The digestive tube and its general functions.

Big Sister loudly exclaimed, "Mom! Baby Heinie just swallowed a marble!" Assuming that the kid didn't choke or aspirate the marble, would Mom ever see it again? The answer, of course, would probably be this: Mom would see the marble in Baby Heinie's diaper, the next morning! The marble, obviously, was defecated or *egested* (**ee-JES**-ted) by Baby Heinie.

Ingestion versus Egestion

Of the major processes associated with the digestive tube, *ingestion* (**in-JES**-chun) is the first, while *egestion* (**ee-JES**-chun) is the last. Ingestion is the "process of carrying (food) into" the digestive tube, while egestion is the "process of carrying (feces) out." In both ingestion and egestion, then, stuff is being carried, either "into" (*in-*) or "out of" (*e-*), the digestive tube.

As far as Baby Heinie goes, he ingested the marble through his oral cavity, then finally egested it through his anus. Now, the marble was essentially in the same condition *after* it was defecated or egested (if Mom cleans it up), as it was *before* it was even ingested, wasn't it? This fact demonstrates an important principle about the digestive tube. To be colorful, we will call it *The "Rabbit-Hole" Principle*:

A

> **THE "RABBIT-HOLE" PRINCIPLE:** *Anatomically speaking, the digestive system is a tube that lies outside the rest of the body.*

Organ System 2

If you doubt the truth of this principle, think about what we just talked about with regards to Baby Heinie. The kid ingested the marble into his digestive tube at one end. The marble passed all the way through the tube. Finally, it was egested (defecated) out the other end of the tube. For all practical matters, then, the marble didn't interact with any other part of the body, did it? Thus, that marble Baby Heinie swallowed might just as well have gone down some real "rabbit hole!" Since the lumen or interior of the digestive tube lies outside of the rest of the body, a marble or some other object can be swallowed and finally passed out in the feces, without having interacted with any other parts of the body!

This principle does have some practical value, when you consider that there are some things that belong *inside* of the digestive tube, but *not* anywhere else in the body! Consider, for example, the bacteria called *Escherichia* (esh-er-**EYK**-ee-ah) *coli* (**KOH**-leye) or *E. coli* for short. (The tongue-twisting first word in the name of this bacteria comes from Theodor Escherich, the

German physician who discovered the organism.) *Escherichia coli* (*E. coli*) is a type of bacteria commonly found within the *colon* (**KOH**-lun) or "large intestine" (*col*) of humans and other animals. As such, they are present in huge quantities within the feces. As long as these bacteria stay inside of the large intestine (colon), they are *nonpathogenic* (**NAHN**-path-oh-**JEN**-ik), or "not disease-causing."

But if they get almost anywhere else in the body, they can be very dangerous! While the inside of the colon is "dirty" (containing feces and millions of bacteria), the outside of the colon (the surrounding abdominopelvic cavity) is very "clean" (free of bacteria and sterile). Say that a person suffers a rupture of the *vermiform* (**VERM**-ih-form) *appendix*, the "wormlike attachment" to the colon. Feces and swarms of *E. coli* pour out of the ruptured colon, and into the formerly sterile abdominopelvic cavity. Severe clinical disorders, such as *peritonitis* (**pair**-ih-ton-**EYE**-tis) can result. Peritonitis is an "inflammation of" (*-itis*) the "peritoneum" (**pair**-ih-ton-**EE**-um), the membrane lining the wall of the abdominopelvic cavity and covering the abdominal viscera. High fever, chills, vomiting, and even shock and heart failure may eventually result!

So, it's a darn good thing that the digestive tube is like an isolated "rabbit hole," isn't it?

Digestion and Absorption

Taking a peek at Figure 14.1, *digestion* and *absorption* are two additional functions carried out within the digestive tube. Digestion literally means a "dividing or dissolving." Digestion is defined, then, as the chemical or physical breakdown ("dividing") of food. In Figure 14.1, a nice sandwich is digested into many smaller pieces.

When a sandwich or any other food is digested, various *nutrients* are released. These include, of course, substances such as glucose, fatty acids, and amino acids, which the body can utilize for energy or for constructing new structures.

A common alternate name for the digestive tube or tract is the *alimentary* (**al**-uh-**MEN**-tur-ee) *canal*. The word, alimentary, "refers to nourishing." The alimentary canal is a "nourishing canal" in that it digests or breaks down larger molecules (such as starch or glycogen) into smaller molecules (such as glucose), which are *nutrients* (substances that provide nutrition).

After ingested foods are digested into smaller nutrients, the nutrient particles are *absorbed*. The term, *absorption*, means the movement of material from the lumen of the digestive tract into the bloodstream. Generally speaking, foods must first be chemically or physically digested into smaller pieces (nutrients) that are ultimately absorbed across the wall of the digestive tract, into the bloodstream, before being circulated on to feed the various body tissues.

Secretion and the Accessory Digestive Organs

The fifth and final function of the digestive (alimentary) tract is *secretion*. We have already encountered secretion, of course, as the major function of glands (Chapter 10). With glands, we defined secretion as the release of useful products. There are, indeed, a number of both endocrine and exocrine glands closely associated with the digestive tube.

Our major focus in this chapter, however, will be upon the secretions of the *accessory digestive organs*, which are added to the digestive tube contents. The accessory digestive organs are organs that are attached to the sides of the digestive tube, but through which no food or feces actually passes.

As shown in Figure 14.1, the accessory digestive organs include the *salivary* (**SAH**-lih-**vair**-ee) *glands*, pancreas, liver, and *gall bladder*. These organs all add small quantities of various secretions to the digestive tube, which help in its process of breaking down big chunks of food into smaller molecules of nutrients.

General Functional Summary

We can list the five general functions of the digestive tube (alimentary canal) as:

$$\begin{array}{ll} \textbf{GENERAL DIGESTIVE} = & \textbf{Ingestion + Digestion + Absorption} \\ \textbf{TRACT FUNCTIONS} & \textbf{+ Secretion + Egestion (Defecation)} \end{array}$$

We will now proceed to dissect the major portions of the digestive tract, moving cranially to caudally. As we go along, their chief characteristics and body functions will also be briefly outlined (Figure 14.2).

Organ System 1

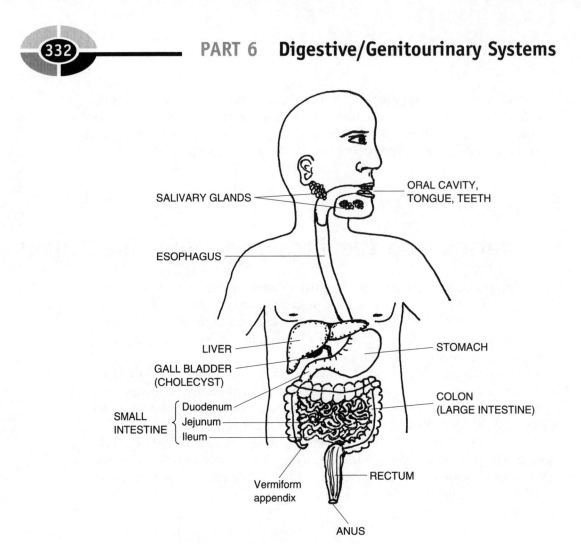

Fig. 14.2 Major features of the human digestive tube.

The Oral Cavity, Pharynx, and Esophagus

Ingested food immediately enters the mouth (oral cavity), where *physical digestion* of all three major types of foodstuffs (carbohydrates, lipids, and proteins) begins. Physical digestion is just the mechanical breaking apart of ingested food, using the teeth, lips, and gums.

Chemical digestion of carbohydrates also begins within the oral cavity. This is due to the presence of "spit" or *saliva* (sah-**LIE**-vah), which is secreted into the mouth by the salivary glands. The saliva contains various digestive enzymes, such as *salivary amylase* (**AM**-ih-**lace**), or "starch" (*amyl*) "splitter." Salivary amylase begins the chemical digestion (breakdown) of *complex* *carbohydrates*, such as starch, into double-sugars. [**Study suggestion:** Eat a plain saltine cracker. At first, it is quite dull, reflecting its complex starch content. But as you chew it and mix it with your saliva, notice that it begins to taste sweet. To what specific chemical should you give credit for this drastic change?]

By the time a person is done chewing food and mixing it with saliva, the general result is a food *bolus* (**BOH**-lus). A food bolus is a soft "ball" (*bol*) of partially digested food.

When done chewing, the person uses the tongue and flips the food bolus into the back of the pharynx. (The throat or pharynx is the passageway shared by both the respiratory and digestive systems.) The bolus pushes the epiglottis shut, then slides down into the *esophagus* (eh-**SAHF**-uh-**gus**) or "gullet," the muscular tube leading into the stomach.

The upper portion of the esophagus is lined by voluntary striated (cross-striped) muscle. Hence, the first part of swallowing is voluntary. ("So, why did I just gulp down that piece of delicious apple?" you might well ask yourself. "Because I darn well *wanted* to, that's why!")

However, the lower 2/3 of the esophagus is lined by mostly smooth, involuntary muscle. This smooth muscle layer is found in a part of the digestive tube wall called the *muscularis* (mus-kyoo-**LAY**-ris) (see Figure 14.3).

Because the lower esophagus is lined by smooth muscle within the muscularis, the latter part of swallowing is *not* under our conscious control. Therefore, once a swallowed food bolus has entered the lower esophagus, you just have to let it go down into your stomach! ("Oh, oh!" you might suddenly question yourself. "Didn't I just see half a worm in that chunk of apple?" Too late! You've *already* swallowed it! You can't back out, now!)

The Four Basic Tunics in the Digestive Tube Wall

From the lower 2/3 of the esophagus, all the way down to the *anal canal* (passageway leading to the anus), the digestive tract has the same four basic *tunics or layers* in its wall. [**Study suggestion:** Can you remember the three tunicas or coats in the blood vessel wall? Try to name them, now. Do you remember the basic structure and function of each tunica? If need be, review Chapter 11. While reading the rest of this chapter, keep asking yourself, "How do the *4 tunics* in the *digestive tube wall* compare to the *3 tunicas* within the *blood vessel wall?*]

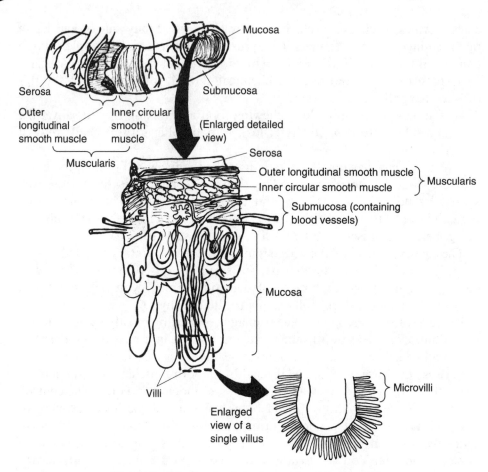

Fig. 14.3 The four basic tunics on the digestive tube.

TUNIC #1: THE MUCOSA

Lying in direct contact with the digestive tube lumen is the *mucosa* (mew-**KOH**-sah) (see Figure 14.3). The mucosa is the innermost "mucous" or "slime" (*mucos*)-producing membrane of the digestive tube wall. It is quite similar to the mucous membrane lining many passages of the respiratory pathway (Chapter 13). The thick, slimy mucus it secretes serves to keep the inner tube wall from getting dehydrated (dried out). It also lubricates food or feces as they pass through the tube lumen.

A major functional adaptation of the mucosa is a throwing of its flat membrane surface into raised folds. With a flat membrane, there is only one surface available for the absorption of nutrients. But with a fold, there are three surfaces – the right side, the top, and the left side of each fold. Thus, folding of the mucosa greatly increases the surface area available for the mucosa's major task of absorbing nutrients from the digestive tube lumen.

TUNIC #2: THE SUBMUCOSA

Present immediately "below" (*sub-*) the mucosa, one finds the *submucosa* (sub-mew-**KOH**-sah). The submucosa is a dense fibrous connective tissue with a rich supply of both blood vessels and elastic fibers. The elastic fibers allow digestive tube structures to stretch when they contain food or feces, and then snap back to normal size after material has passed. It also is the place where a lot of the nutrients being absorbed through the mucosa wind up. From here, the vessels of the submucosa branch out and supply nutrients to the rest of the digestive tube wall.

TUNIC #3: THE MUSCULARIS

The muscularis is the smooth muscle portion of the digestive tube wall. It has two main components – an inner *circular smooth muscle layer* plus an outer *longitudinal smooth muscle layer*. The outer longitudinal smooth muscle layer runs "lengthwise" (*longitudinally*) down the digestive tract. Its main job is exciting the circular smooth muscle layer to contract. The inner circular smooth muscle layer is like a noose around a neck. When it contracts, it constricts the digestive tube lumen, thereby pushing food or feces along. The inner circular smooth muscle, therefore, is primarily responsible for digestive tube movements. In a number of places within the tube, the circular smooth muscle also thickens and develops into *anatomical sphincters* (**SFINGK**-ters), which can strongly close off the tube lumen and prevent flow, at times.

Capsulizing:

Tissue 1

THE MUSCULARIS = OUTER LONGITUDINAL + INNER CIRCULAR
(*Smooth muscle layer*)   SMOOTH MUSCLE   SMOOTH MUSCLE
(*Functions to excite circular   (Functions to constrict*
smooth muscle layer*)   *tube lumen and cause
***tube movements*)**

TUNIC #4: THE SEROSA

The fourth and most peripheral tunic covering the digestive tube is called the *serosa* (**see-ROH**-sah). The serosa is a type of *serous* (**SEER**-us) membrane, one that secretes a "watery" (*ser*) fluid which moistens the outer surface of the digestive tube. Its other name is the *visceral peritoneum* (**per**-ih-toh-**NEE**-um). This alternate indicates that the serosa is also the serous membrane that covers the viscera (internal organs) within the abdominopelvic cavity. Since it is the outermost covering, the serosa (visceral peritoneum) also helps anchor the digestive tube within the cavity, with its many strong collagen fibers.

A **SUMMARY**

We can summarize the above information:

Organ System 3

THE FOUR BASIC	**= MUCOSA + SUBMUCOSA**
TUNICS ON THE TUBE	**+ MUSCULARIS + SEROSA**

The Stomach: A J-shaped Pouch

The stomach is a capital J-shaped pouch that acts as a temporary storage place for ingested food. As seen in Figure 14.4, there is a *greater curvature* sweeping along the inferior edge of the stomach, as well as a *lesser curvature* arcing along its superior edge. Attached along the greater curvature, one finds the *greater omentum* (oh-**MEN**-tum). The greater omentum is a fatty "covering" (*oment*) that is actually a double-fold of visceral peritoneum. It hangs down like a fatty apron from the greater curvature, covering much of

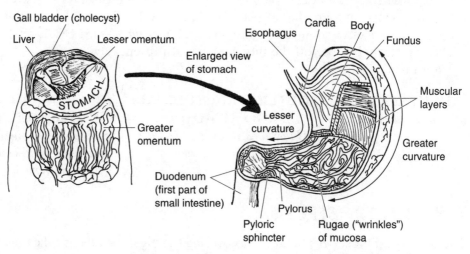

Fig. 14.4 The stomach and its parts.

the colon and small intestine. The *lesser omentum* attaches along the lesser curvature at the top of the stomach, and helps anchor the stomach to the liver and spleen.

MAJOR REGIONS OF THE STOMACH

Major regions of the stomach include the *cardia* (**KAR**-dee-ah), *body*, *fundus* (**FUN**-dus), and *pylorus* (pie-**LOR**-us). The cardia is the superior opening of the stomach, where the esophagus empties into it. The cardia (like the lower esophagus) lies very near the "heart" (*cardi*), hence its name.

The fundus is the rounded, dome-shaped bulge lying just lateral to the cardia. The body, situated just below the fundus, is the large midportion of the stomach. Finally, the pylorus or "gatekeeper" is a small room-like pouch at the distal end of the stomach.

ACTIONS AND LINING OF THE STOMACH

The epithelial cells lining the "stomach" (*gastr*) secrete the *gastric* (**GAS**-trik) *juice*. The gastric juice is especially rich in *hydrochloric* (**HIGH**-droh-**klor**-ik) *acid*, abbreviated as *HCl*, and *pepsin* (**PEP**-sin). Being an extremely strong acid, HCl breaks down rapidly and donates lots of H^+ ions, making it highly reactive and corrosive. Thus, hydrochloric acid begins the chemical digestion of lipids and proteins, as well as continuing the digestion of carbohydrates. Pepsin is an enzyme that helps break down proteins, as well.

The mucosa in the stomach is thrown into a number of long *rugae* (**ROO**-guy) or "wrinkles." These rugae increase the surface area for quick absorption of salt, water, and alcohol across the stomach wall.

Due to the action of the gastric juice, the food bolus from the esophagus is now changed into *chyme* (**KIGHM**). The chyme is a thick, soupy mass of partially digested material. Since it is almost a liquid, chyme is like a "juice" (*chym*) that leaves the stomach through a muscular ring called the *pyloric* (pie-**LOR**-ik) *sphincter*. [**Study suggestion:** If the whole pylorus region is literally the "gatekeeper" to the small intestine, then the pyloric *sphincter*, alone, must be the "_____."]

"If the stomach is full of so much HCl, then why doesn't it digest itself?" Part of the answer is that the stomach secretes a highly *alkaline* (**AL**-kah-**lin**) or basic layer of mucus. This mucus is a protective "slime" (*muc*) about 1 mm thick, that effectively coats the stomach lining and neutralizes most of the acid that contacts it.

Acid mixed with chyme pushes the pyloric sphincter open, so that it enters the small intestine.

A

SUMMARY

The major regions of the stomach can now be put into a summary form:

Organ 1

**MAJOR STOMACH = CARDIA + FUNDUS + BODY + PYLORUS
REGIONS**

The Small Intestine = Twenty Twisted Feet of Wondrous Tubing!

The *small intestine* is a small-diameter, extensively folded tube that averages about 20 feet (6 meters) in length. We have called it a "wondrous" tubing, because of all the wonderful and amazing feats of digestion and absorption that occur within its 20 feet of tubing.

WE HAVE "12" FINGERS IN OUR DUODENUM!

The first segment of the small intestine is called the *duodenum* (dew-**AH**-den-um), from the Medieval Latin for "presence of 12." Thus, the length of the duodenum was measured by ancient anatomists as 12 finger-breadths, placed side-by-side. Figure 14.5 shows this, and also demonstrates that this first "12 finger-breadths" of the small intestine is a common meeting ground for chyme and various digestive secretions.

The duodenum receives chyme from the pylorus of the stomach, and also secretions from three of the accessory digestive organs (the liver, gall bladder, and pancreas). The liver is a large, brown, multi-lobed organ that produces and secretes *bile*, as well as many other useful substances. Bile is a brownish-green detergent substance that *emulsifies* (ih-**MUL**-sih-feyes) fat within the small intestine. *Emulsification* (ih-**mul**-suh-fuh-**KAY**-shun) is literally the "process of" (-*tion*) "milking out" (*emulsif*) one non-mixable fluid substance from another one. Consider, in this case, the large globules of partially digested fat that do not mix very well with the soupy chyme entering the duodenum from the stomach. Bile from the liver acts to emulsify the large fat globules, breaking them apart (or, in a sense, "milking them out" of the rest of the chyme). As a result, a separate foam of tiny fat droplets is created within the small intestine. [**Study suggestion:** Pour some liquid detergent onto

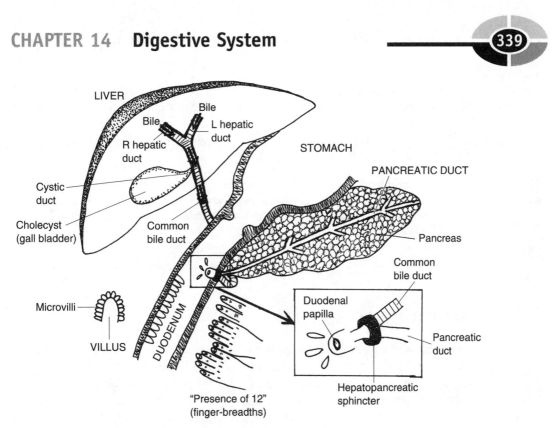

Fig. 14.5 The duodenum and its friendly neighbors.

a bunch of greasy plates, and then observe what happens. In what way does this liquid detergent act somewhat like bile?]

A MEETING WITH THE GALL BLADDER

Bile is secreted continuously, day and night, into the right and left *hepatic* (heh-**PAT**-ik) or "liver" *ducts*. These ducts carry the bile into the *cystic* (**SIS**-tik) *duct*. Cyst means "bladder" or "sac," while *chole* (**KOH**-lee) is Latin for "bile or gall." Hence, the compound word, *cholecyst* (**KOH**-luh-sist), translates into English as "gall bladder" or "bile sac."

The cholecyst (gall bladder) is a muscular-walled sac that receives bile from the liver and stores it temporarily. When the duodenum becomes swollen with fatty chyme, a hormone is released that stimulates the walls of the gall bladder to contract. A load of bile is squirted out of the cholecyst, much like a slug of brownish-green pea soup or gravy being squeezed out of a rubber balloon.

The bile squirts into the cystic duct, and then into the *common bile duct*, which carries it the rest of the way down into the duodenum. Here, then, the bile triggers the emulsification of fat.

To capsulize the information on bile, remember *THE WONDERFUL LIVER BILE-BLADDER RULE:*

Organ 1

> *"Bile is continuously produced and secreted by the liver, but is then temporarily stored and released into the duodenum by the cholecyst (gall bladder)."*

THE PANCREAS BUTTS IN

As pictured in Figure 14.5, the *pancreatic* (pan-kree-**AT**-ik) *duct* extends from the pancreas and merges with the base of the common bile duct. The pancreatic duct is the main passageway for the *pancreatic juice*. Surrounding both of them at their point of union is the *hepatopancreatic* (heh-**PAT**-oh-pan-kree-**AT**-ik) *sphincter*. The hepatopancreatic sphincter is a ring of smooth muscle that regulates the emptying of both the common bile duct (the hepatic or "liver" portion) and the pancreatic duct into the small intestine.

When this sphincter (muscular ring) relaxes, bile and pancreatic juice flow through the *duodenal* (dew-**AH**-deh-nal) *papilla* (pah-**PIL**-lah). The duodenal papilla is a "little nipple or pimple" (*papill*)-like projection with a hole in its center. Bile from the liver, as well as pancreatic juice from the exocrine gland portion of the pancreas (Chapter 10), drip into the duodenum through the hole in the duodenal papilla.

The pancreatic juice contains *sodium bicarbonate* (buy-**KAR**-buh-**nayt**), symbolized chemically as $NaHCO_3$, as well as a variety of digestive enzymes. These enzymes include amylases (starch-splitters), *lipases* (**LIE**-pay-sez) or "fat-splitters," and *proteases* (**PROH**-tee-**ay**-sez) or "protein-splitters." The lipases, for example, complete the chemical digestion of fat or lipids, after they have been emulsified by bile into a fatty foam. The resulting products, such as *fatty acids* and the substance *glycerol* (**GLIH**-sir-**ahl**), are then absorbed across the walls of the small intestine, and into the bloodstream.

Similarly, the proteases continue the chemical breakdown of proteins into amino acids, which are also absorbed into the bloodstream. And the amylases in the small intestine generally finish the chemical breakdown of carbohydrates into simple sugars such as glucose, which are then absorbed.

Capsulizing the small intestine's absorption of nutrients:

NUTRIENTS	*Simple sugars*	+	*Amino acids*	+	**Fatty acids**
ABSORBED	*(such as glucose)*		*(from proteins)*		**& glycerol**
IN THE SMALL					*(from fats)*
INTESTINE:					

Organ 2

VILLI AND MICROVILLI: OUR LITTLE "TUFT" GUYS

Reflecting its critical role in the absorption of nutrients, there are several important modifications to the portion of the mucosa lining the small intestine. These modifications occur within the mucosa of the duodenum, as well as in the mucosa of the next two portions of the small intestine – the *jejunum* (jeh-**JOO**-num) and the *ileum* (**IL**-ee-um).

The mucosa of all three portions of the small intestine (duodenum, jejunum, and ileum) is thrown into thousands of *villi* (**VIL**-ee). Review of Figure 14.5 reveals that each single *villus* (**VIL**-us) resembles a little bump or curved "tuft of hair" (*vill*). The surface of each villus, in turn, is covered with dozens of *microvilli* (**MY**-kroh-**vil**-ee).

The microvilli are little bumps upon each villus ("tuft of hair"). This peculiar pattern makes them look like many really "tiny" (*micro-*) "tufts of hair" (*villi*). The numerous villi (as well as the microvilli around the edge of each villus) throw the mucosa up into hundreds of tiny bumps. Each tiny bump having three sides (top, right side, and left side), the amount of surface area available for absorption is vastly increased! Hence, the absorption of nutrients into the bloodstream of the submucosa, from the lumen of the small intestine, is extremely efficient.

SMALL INTESTINE SUMMARY

Summarizing all of the above, we can say that the duodenum, as the first segment of the small intestine, receives it all! That is, the duodenum gets chyme from the stomach, bile from the liver and cholecyst, and pancreatic juice from the pancreas. As a result, the chemical digestion of all three basic types of foodstuffs – carbohydrates, lipids, and proteins – is essentially completed within the small intestine.

Finally, lying downstream from the duodenal papilla (the combined liver and pancreas entry point), one finds the jejunum and ileum. The ancient anatomists usually found the lumen of the jejunum to be "empty" (*jejun*) when they cut it open. But that's quite a bit of "emptiness," since the jejunum is about 8 feet (or 2.4 meters) in length!

The last portion of the small intestine is the ileum, which is located at about the level of our body "flanks" (*ile*). (Remember that the flank is the portion of the back lying below the ribs, but above the top of the *ilium* or hip bone.) The ileum is lighter in color, slightly narrower in diameter, and has a thinner wall, compared to the jejunum. The jejunum also has *circular folds* within its lining, and also larger villi than the ileum.

The jejunum and ileum basically complete the processes of chemical digestion and absorption of nutrients that began in the duodenum.

We end our discussion of the small intestine with the following structural summary:

SMALL INTESTINE = DUODENUM + JEJUNUM + ILEUM

Organ 2

The Colon: Our Large Intestine Brings Up the Rear

The last major section of the digestive tube is the *colon* (**KOH**-lun) or "large intestine." The colon (large intestine) is a wide-diameter, folded tube, about 6 feet (2 meters) in length in an average-sized adult (see Figure 14.6).

The *ileum*, being the *last* portion of the *small intestine*, empties its fluid chyme into the *cecum* (**SEE**-kum), the *first* part of the *colon*. The cecum is a "blind" (*cec*) or dead-ended pouch that has the vermiform appendix hooked to its base. The vermiform appendix is basically a solid attachment of modified lymphatic tissue that plays a minor role in the body's immune or self-defense system.

The liquid chyme pushes from the ileum, and through the *ileo-cecal* (**il**-ee-oh-**SEE**-kul) *sphincter*. This is an anatomical sphincter consisting of a ring of circular smooth muscle that opens into the cecum. Once within the cecum, the chyme begins to undergo an extensive drying out process. Large amounts of salt and water are absorbed. In addition, there are beneficial bacteria in the colon that produce a variety of B-vitamins, as well as sulfur-containing amino acids, which are also absorbed.

Due to this drying out process, chyme is modified into feces within the colon. Besides H_2O, feces also contain a significant percentage of *fecal* (**FEE**-kal) *bacteria* and *dietary fiber* (actually undigested plant cell wall material).

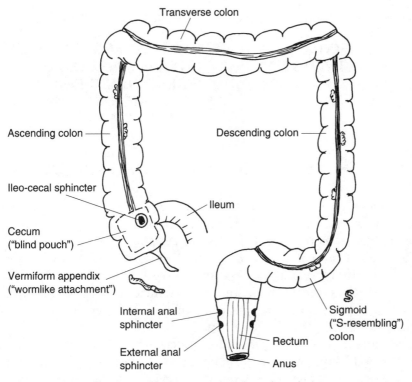

Fig. 14.6 The human colon (large intestine).

After the cecum, next in sequence are the *ascending colon* (which goes upward), *transverse colon* (which runs horizontally), and the *descending colon* (which goes downward). *Sigmoid* (**SIG**-moyd) means "S-resembling." Hence, the *sigmoid colon* is the S-resembling portion of the large intestine coming right after the descending colon.

The sigmoid colon snakes down into the *rectum* (**REK**-tum). The rectum is a "straight" (*rect*) muscular-walled tube that empties feces into the anus. There are two sphincters within the rectum. The higher one, called the *internal anal sphincter*, is not under our conscious control. The internal anal sphincter opens automatically whenever some feces have moved down from the sigmoid colon and into the upper portion of the rectum.

Fortunately for us, there is also a ring of voluntary striated (cross-striped) muscle, positioned in the lower portion of the rectum. This muscular ring is called the *external anal sphincter*. Its contraction and relaxation is very much under our conscious control (at least, ever since we were first "potty-trained")! Therefore, we can usually choose the time and place where we will consciously relax this lower sphincter and carry out defecation (egestion).

Organ 3

Let us bring up the rear by summarizing the parts of the colon:

$$\begin{aligned}
\textbf{THE COLON} \quad &= \textbf{CECUM + ASCENDING COLON} \\
\textbf{(LARGE INTESTINE)} \quad &\quad \textbf{+ TRANSVERSE COLON} \\
&\quad \textbf{+ DESCENDING COLON} \\
&\quad \textbf{+ SIGMOID COLON} \\
&\quad \textbf{+ RECTUM + ANUS}
\end{aligned}$$

Colon Polyps: A Warning Sign for Cancer

Tissue 1

We have learned that the mucosa has various modifications that enhance absorption of nutrients across it. The mucosa of the colon, however, has some modifications that are definitely *not* health-enhancing!

These modifications are *colon polyps* (**PAHL**-ips). A polyp is literally a little "foot." It is a small, "foot"-shaped growth that projects from the surface of a mucous membrane. Although the mucosa of the large intestine is not the only place where polyps can occur, it is a very dangerous place, because they often grow for long periods, undetected.

A potentially life-saving procedure to detect colon polyps is *colonoscopy* (**koh**-lahn-**AHS**-koh-pee) – an "examination of" (-*oscopy*) the "colon" with a lighted instrument. The *colonoscope* (koh-**LAHN**-oh-skohp) can be inserted far into the rectum, all the way up into the cecum! The examining physician can visually detect colon polyps at high magnification, and see them projecting like shiny little feet from the colon mucosa.

The most common type of polyps are *hyperplastic* (**high**-per-**PLAS**-tik) *polyps*. These are *benign* (beh-**NINE**) – "kind" or non-cancerous polyps – that simply represent an "excessive or above normal" (*hyper*-) "formation" (-*plasia*) of epithelial cells. This type never evolves into colon cancer, but can cause *rectal bleeding* if they become too large.

The polyps that are dangerous are called *adenomatous* (**ad**-eh-**NOH**-mah-tus) *polyps*. These are literally "gland" (*aden*) "tumors" (-*omas*) that are shaped like little "feet" (*polyps*). Although the adenomatous polyps themselves, are benign, they are considered *precancerous* polyps that, if left to grow long enough, can develop into full-blown *cancer of the colon*.

One type of precancerous polyp is the *villotubular* (**vil**-oh-**TOOB**-you-lar) *adenoma* (**ad**-eh-**NOH**-mah). When a villotubular adenomatous polyp is removed during colonoscopy and examined under a microscope, a highly abnormal pattern of *cellular architecture* (cellular arrangement) is observed. Specifically, the tissue sample has fine, "hair" (*villo*-) projections sticking out of its mass of epithelial cells. And in between these hair-like villi, are hollow "little tubes" (*tubul*). This highly unusual mixture of abnormal cellular patterns is the probable forerunner of colon cancer.

Everyone is recommended to have a routine *screening colonoscopy* by age 50, or much earlier, if a history of colon cancer runs in the family, or if rectal bleeding has been experienced.

Quiz

Refer to the text in this chapter if necessary. A good score is at least 8 correct answers out of these 10 questions. The answers are listed in the back of the book.

1. The digestive tube is located:
    (a) Between the oral cavity and the diaphragm
    (b) Along the peripheral veins and arteries, only
    (c) From the oral cavity to and including the anus
    (d) Around the pharynx and gullet

2. The digestive system can be thought of as "rabbit hole" lying outside of the body, because:
    (a) Everything that is ingested is immediately defecated
    (b) Some matter may be ingested and passed in the feces, without ever having interacted with the rest of the body
    (c) Some particles are secreted, while others are absorbed
    (d) All possible nutrient sources are thoroughly digested and absorbed

3. The salivary glands, pancreas, liver, and gall bladder:
    (a) Essential parts of the digestive tract
    (b) Endocrine glands secreting hormones that affect the digestive processes
    (c) Release lots of acids and enzymes
    (d) Accessory digestive organs

4. After you have swallowed a rotten apple filled with worms into your lower esophagus:
    (a) Spit it out!
    (b) You have to finish swallowing it!
    (c) Just stop, and think it over
    (d) Call Baby Heinie's Mom!

5. The layer of the digestive tube tunic that consists of smooth muscle tissue:
   (a) Muscularis
   (b) Mucosa
   (c) Serosa
   (d) Submucosa

6. A double-fold of visceral peritoneum attached along the inferior margin of the stomach:
   (a) Lesser curvature
   (b) Polyposis
   (c) Greater omentum
   (d) Fatty ejaculate

7. Superior portion of the stomach lying below entrance of the esophagus:
   (a) Jejunum
   (b) Fundus
   (c) Cardia
   (d) Pylorus

8. The gall bladder is alternately called the:
   (a) Serosa
   (b) Duodenum
   (c) Cholecyst
   (d) Biliary tubule

9. The hepatopancreatic sphincter serves to:
   (a) Allow HCl to freely enter the esophagus
   (b) Release the bile and pancreatic juice into the duodenum
   (c) Permit pancreatic juice to back-up into the salivary glands
   (d) Pour strong HCl directly onto the surface of both lungs

10. The colon begins with this structure:
    (a) Cecum
    (b) Vermiform appendix
    (c) Transverse colon
    (d) Sigmoid colon

Body-Level Grids for Chapter 14

Several key body facts were tagged with numbered icons in the page margins of this chapter. Write a short summary of each of these key facts into a numbered cell or box within the appropriate *Body-Level Grid* that appears below.

Anatomy and *Biological Order* Fact Grids for Chapter 14:

A

ORGAN
Level

1	2

3

ORGAN SYSTEM
Level

1	2

3

Physiology and *Biological Order* Fact Grids for Chapter 14:

TISSUE
Level

1

ORGAN
Level

1	2

ORGAN SYSTEM
Level

1

Anatomy and *Biological Disorder* Fact Grid for Chapter 14:

TISSUE
Level

1

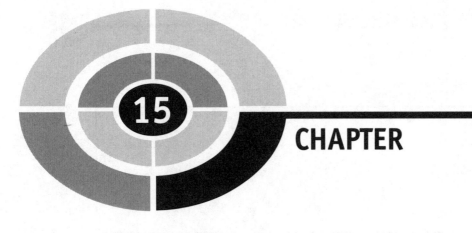

The Genitourinary System – Keeper of Our Urine and (GASP!) Sex

Well, here we are, Baby Heinie! We have *arrived*! "Arrived at *what*, Professor?" We have arrived at Chapter 15, the *last* chapter in *ANATOMY DEMYSTIFIED*! But, unlike the first 14 chapters, I'm afraid that you might have to *sit this one out*, my mischievous little friend! "*Why* is that, Prof?"

Well, we are going to discuss the *genitourinary* (**JEN**-ih-toh-**ur**-ih-**nair**-ee) *system!* It is also called (when you reverse the two word roots), the *urogenital* (**you**-roh-**JEN**-ih-tal) *system*. And for your information, Baby Heinie, the *genital* (**JEN**-ih-tal) *organs* are the ones we adults use to "beget or produce" (*genit*) sexually.

"Oh, boy! You mean I get to learn about *sex*?" Well, if your Mom and Dad approve, I think it'll be okay! You know, it's important for young people to know *accurate information* about their genitals (sex organs), rather than learning about it from their giggling and snickering friends! So, *away* we go!

The Genitourinary (Urogenital) System Concept

In anatomy, we frequently hear talk of a single *urinary* (**YOUR**-ih-**nair**-ee) *system*. This urinary system literally "pertains to" *(-ary)* "urine" production, storage, and excretion from the body. And we also hear about a single *reproductive system* in both the male and female. The reproductive system is exactly about "producing" a new human being, "again" (*re-*).

However, it is really more appropriate to speak not of just the urinary and reproductive systems alone but of a combined *genitourinary* or *urogenital* *system*. The reason for this is that many of the structures of the urinary and reproductive (genital) organs are *shared in common*. Consider, for example, the *penis* (**PEA**-nis) in males. The penis is a spongy "tail" (*pen*)-like organ that serves both to carry urine out of the body, as well as deliver *spermatozoa* (**sper**-mat-oh-**ZOH**-ah) – the "seed" (*spermat*) "animals" (*zo*). (After all, the spermatozoa or *sperm cells* really do look a lot like wriggling little tadpoles!)

Hence, we state this basic equation:

GENITOURINARY =	GENITAL OR	+ URINARY
(UROGENITAL)	REPRODUCTIVE	ORGANS
SYSTEM	ORGANS	

A

Organ System 1

The Kidney and Its Connections

The major organs of urine excretion are the *kidneys*. Our kidneys are a pair of reddish-brown, bean-shaped organs flanking either side of the vertebral column, very deep within the back.

RENAL (KIDNEY) ANATOMY

Figure 15.1 provides an overview of *renal* (**REE**-nal) or "pertaining to" (*-al*) "kidney" (*ren*) anatomy. The kidney is enclosed within the *renal capsule*, which is a thin membrane consisting of fibrous connective tissue. The kidney, itself, is subdivided into three major areas or zones.

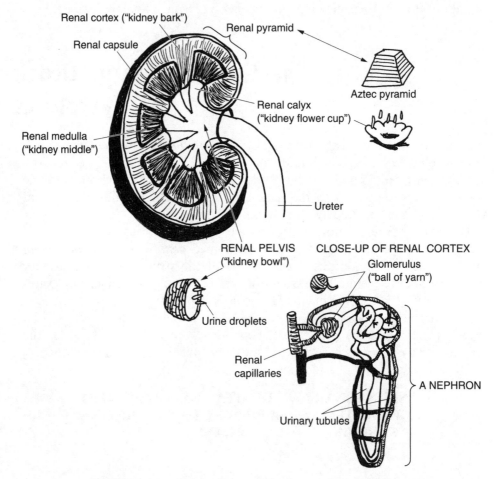

Fig. 15.1   An overview of gross and microscopic renal anatomy.

The outermost zone is called the *renal cortex*. Much as the adrenal cortex forms a thin "bark" over the surface of the adrenal body (Chapter 10), the renal cortex does the same for the kidney. The "middle" (*medull*) area is the *renal medulla* (meh-**DEW**-lah). And the deepest zone is the *renal pelvis* (**PEL**-vis). The renal pelvis is a broad, bowl-shaped sac that receives the urine as it flows from the renal cortex and medulla. And carrying the collected urine out of the renal pelvis is the *ureter* (**YOUR-eh-ter**).

Capsulizing the above, we obtain:

$$\text{THE 3 MAJOR ZONES} = \text{RENAL} + \text{RENAL} + \text{RENAL}$$
$$\text{OF THE KIDNEY} \qquad \text{CORTEX} \quad \text{MEDULLA} \quad \text{PELVIS}$$

MICROSCOPIC ANATOMY OF THE KIDNEY

Organ 1

So far, we have provided an overview of gross renal anatomy. But the real "business" of urine formation goes on within the microscopic anatomy of each kidney!

Millions of *nephrons* (**NEF**-rahns) are scattered throughout the renal cortex. Let's state ***THE HARD-WORKING NEPHRON PRINCIPLE:***

> *The nephrons are the major microscopic functional units of the kidneys.*

The nephrons *have* to be hard-working, because they are the structures actually responsible for the formation of urine from our blood.

Each nephron begins with a *glomerulus* (gluh-**MAHR**-yew-**lus**). The glomerulus is a tiny, red-colored collection of *renal capillaries*. This structure gets its name from its resemblance to a little red "ball of yarn" (*glomerul*). The blood pressure pushing against the walls of the capillaries in each glomerulus causes a filtration of fluid out of the glomerulus and into the adjoining group of *urinary tubules* (**TWO**-byools) – "tiny urine tubes."

The urinary tubules from each group of neighboring nephrons eventually empty into a common passageway called a *collecting duct*. A number of collecting ducts pass down together through the renal medulla. They create the *renal pyramids*, which are pointed at their bottom tips like the rather blunt pyramids constructed by the Aztecs or Inca Indians. The tip of each renal pyramid drips urine into a *renal calyx* (**KAY**-licks), or "kidney flower cup." And the urine from each calyx eventually flows into the body of the renal pelvis, before it leaves the kidney through the ureter.

The Urinary Pathway

Figure 15.2 reveals the rest of the urinary pathway, lying beyond the kidney. The right and left ureters both dump urine into the *urocyst* (**YUR**-oh-**sist**) or

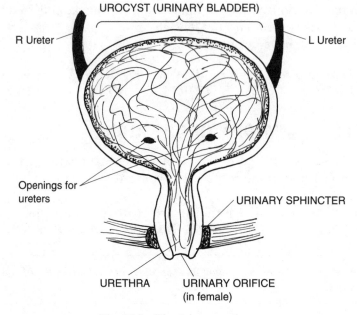

Fig. 15.2   The urinary pathway.

urinary "bladder" (*cyst*). The urocyst (urinary bladder) is a hollow, muscular-walled pouch that temporarily stores the urine before it is excreted.

The urocyst empties into the *urethra* (you-**REETH**-rah), the tube that helps a person literally "make water" (*urethr*) – that is, *urinate* (**YUR**-ih-**nayt**). Surrounding the upper neck of the urethra is the *urinary sphincter*. Much like the external anal sphincter in the digestive pathway (Chapter 14), the urinary sphincter is a ring of voluntary striated muscle. This means, of course, that the contraction and relaxation of this sphincter is under our voluntary control. Thus, after we have been adequately "potty-trained" during early childhood, we can voluntarily relax the urinary sphincter whenever the time and place are right for urination!

At last, urine exits out of the body through the *urinary orifice* (**OR**-ih-**fis**), a tiny, "mouth" (*or*)-like opening.

External Genitals in the Male and Female

Both males and females, of course, have the same basic urinary anatomy and physiology. However, the organs that actually excrete the urine are different, and the organs that surround the external *genital or reproductive openings* in both sexes, are somewhat different.

THE CONCEPT OF SEXUAL HOMOLOGUES

Please note how we just said that they were "somewhat" different. Now, you may recall that in the word, homeostasis, the root, homeo, means "same." A related root, *homo-* (**HOH**-moh), also means "same." From this foundation, we derive a new word, *homologue* (**HAHM**-uh-log). *Sexual homologues* are reproductive structures that have basically the "same" (*homo-*) "relations" (*logue*) in both male and female bodies. In other words, sexual homologues are reproductive organs in the two different sexes that are basically comparable or equivalent in many of their features.

IT ALL STARTS IN THE "SWELLER"

The *external genital organs* of adult males and females are certainly "different," or fully *differentiated* (**dih**-fer-**EN**-she-ay-ted) from one another, aren't they? Such is not the case, however, within the early *embryo* (**EM**-bree-oh). The embryo is literally a "sweller." It represents the first 3 months of life after fertilization. And it is during this critical stage of early life that the embryo's primitive body tissues begin to "swell" (grow in mass due to cell mitosis) and differentiate into more specialized anatomic forms.

There is an *undifferentiated* (**UN**-dih-fer-**EN**-she-ay-ted) or "non-different" early development of the external genitals during the first 2 months (about 8 weeks) of human development. Figure 15.3 (A), for instance, shows a *genital tubercle* (**TOO**-ber-kl) or "little swelling" of tissue that projects from the genital area of a 5-week-old embryo. This genital tubercle always appears the same, even in both sexes!

Derivatives of the glans

At the top of the genital tubercle, we find the *glans* (**GLANS**), which looks much like a rounded "acorn." As development proceeds, the complex program of tissue differentiation quickly unfolds (Figure 15.3, B). By 10 weeks of age the primitive glans of the genital tubercle has already become the *glans penis* in males, but the whole *clitoris* (**KLIT**-or-is) in females. The formation of a glans penis, instead of a clitoris, is largely due to the secretion of two hormones in the male embryo: *testosterone* (tes-**TAHS**-ter-ohn) and the

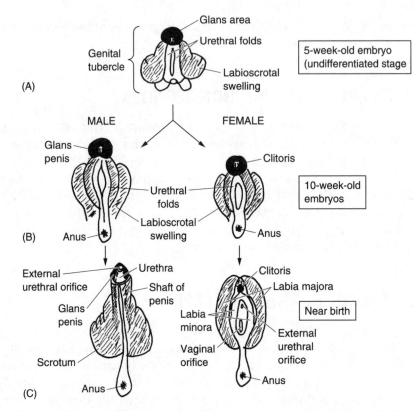

Fig. 15.3   Development of the external genital organs in males and females. (A) Everything starts with the genital tubercle. (B) Formation of glans penis and clitoris. (C) Near-birth genital development.

closely related hormone, *dihydrotestosterone* (die-**HIGH**-droh-tes-**TAHS**-ter-ohn), or *DHT* for short.

Derivatives of the labioscrotal swelling

Inferior and lateral to the glans area in the undifferentiated 5-week-old embryo, we find the *labioscrotal* (**LAY**-be-oh-**SKROH**-tal) *swelling*. By the time of birth (Figure 15.3, C), the labioscrotal swelling has differentiated into the *labia* (**LAY**-bee-ah) *majora* (mah-**JOR**-ah) or outer "major lips" of skin around the *vaginal* (**VAH**-jih-nal) *orifice* (**OR**-ih-**fis**) – the "opening of the vagina." In males, however, the labioscrotal swelling differentiates into the long *shaft of the penis* and the *scrotum* (**SKROH**-tum) or *scrotal* (**SKROH**-tal) *sac*.

Derivatives of the urethral folds

The third major part of the genital tubercle in the 5-week-old embryo comprises the *urethral* (you-**REETH**-ral) *folds*. In males, testosterone and DHT stimulate closure of the urethral folds after 10 weeks, and fusion into a single *urethra* (you-**REETH**-rah). Remember that the urethra is literally the urine-carrying tube within the penis that helps the male "make water" (excrete urine from the body). The urethra opens as the *external urethral or urinary orifice*, right at the tip of the penis.

In females, without testosterone or DHT stimulation, the urethral folds remain open and become the *labia* (**LAY**-bee-ah) *minora* (min-**OR**-ah). These are the inner "minor lips" of skin around the vaginal orifice. The urethra becomes a separate urine-carrying tube, and it also has a separate opening, the *external urethral or urinary orifice*. This orifice (as in the male) excretes urine. But it is located just inferior to the clitoris.

SEXUAL HOMOLOGUE SUMMARY

Table 15.1 provides a convenient summary of the development of the major external genital organs in the male and female. It also points out which organs among these are sexual homologues (fairly equivalent body structures).

Table 15.1   External genital organs in males and females, and their sexual homologues.

Original regions within the genital *tubercle*	Genital organs after birth (*sexual homologues*)	
	Male structures	*Female structures*
The glans area	Glans penis	Clitoris
The labioscrotal swelling	Shaft of the penis & scrotum (scrotal sac)	Labia majora
The urethral folds	Urethra within penis	Labia minora

Internal Anatomy & Physiology of the Male Reproductive Pathway

The basic internal anatomy of the male reproductive pathway is displayed in Figure 15.4. The scrotum (scrotal sac) is literally a "leathery bag of skin" that

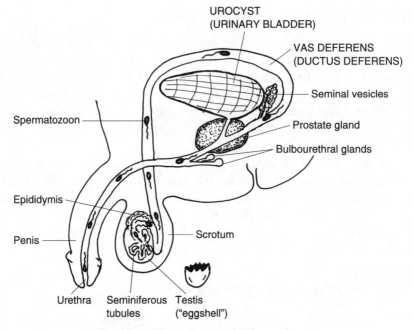

Fig. 15.4   The male reproductive pathway.

suspends the two *testes* (**TES**-teez) outside of the abdominal cavity. Each *testis* (**TES**-tis) is a rather oval, whitish, "eggshell" (*test*)-like structure that contains the *seminiferous* (sem-ih-**NIF**-er-us) *tubules*. The seminiferous tubules are a collection of tiny, highly coiled tubes that carry out the process of *spermatogenesis* (sper-mat-uh-**JEN**-eh-sis). This exactly translates to mean the "production of" (-*genesis*) "sperm" (*spermat*) cells.

From the time of puberty (age 12–13 years) onward, mature sperm cells are continually produced by a *germinal* (**JER**-muh-nal) *epithelium*, which is located in the thick walls of the seminiferous tubules. This germinal epithelium undergoes a constant process of "sprouting" (*germin*) new sperm cells by mitosis, followed by meiosis (a division which reduces the number of chromosomes per cell).

In the human male, each primitive sperm cell contains 46 chromosomes within its nucleus. But after meiosis, the developing sperm cell has this number reduced by 1/2, to a total of just 23 chromosomes. Eventually, a mature *spermatozoon* (sper-**mat**-uh-**ZOH**-un) – sperm cell – with only 23 chromosomes results.

Thousands of mature spermatozoa leave the germinal epithelium of the seminiferous tubules, and are temporarily stored within the *epididymis* (**eh-**

pih-**DID**-ih-mus). The epididymis is a curved, comma-shaped pouch that literally lies "upon" (*epi-*) each testis or "eggshell" (*didym*). The swimming swarms of spermatozoa are ejected from the epididymis during each *ejaculation* (ih-**JACK**-yuh-**lay**-shun). Ejaculation is the "throwing out" (*ejacul*) of *semen* (**SEE**-mun) and spermatozoa from the urinary orifice at the tip of the penis.

Accessory Male Organs Add the Semen

The semen is a thick, milky, sugar-rich, very basic fluid that suspends (floats or holds up) the spermatozoa and gives them nutrition. When the male has an *orgasm* (**OR**-gaz-um), he is literally "swollen and excited." He ejaculates spermatozoa suspended in a fluid of semen. The stored spermatozoa are actively sucked out of the epididymis by strong peristalsis (ring-like muscular contractions) of the walls of the *vas* (**VAHS**) *deferens* (**DEF**-er-enz).

The vas deferens is named for its function as a "carrying away" (*deferens*) "vessel" (*vas*). However, the vas deferens is not really a blood vessel at all. It is alternately called the *ductus* (**DUCK**-tus) *deferens* or "carrying away duct." During male orgasm, the walls of the ductus deferens powerfully and rhythmically constrict or narrow. This negative pressure (suction) event draws the stored spermatozoa out of the epididymis, carrying them over the top of the urinary bladder, and down into the *ejaculatory* (ee-**JACK**-you-lah-**tor**-ee) *duct*. This ejaculatory duct is just a short, elbow-curved linkage to the urethra. The urethra goes all the way through the penis, and ends at the same hole where a man urinates – the external urethral orifice.

Semen is added to the spermatozoa from a number of *accessory male reproductive organs*. These organs include the two *seminal* (**SEM**-ih-nal) *vesicles*, the two *bulbourethral* (**BUL**-boh-you-**REE**-thral) *glands*, and the single *prostate* (**PRAH**-state) *gland*. The seminal vesicles store the "semen" (*semin*). The bulbourethral glands (Figure 15.4) are like two tiny "bulbs" attached to the sides of the "urethra." And the prostate gland is a large, walnut-shaped body that "stands" (*stat*) just "before" (*pro-*) the urethra.

ACCESSORY ORGAN SUMMARY

Capsulizing the above info, we can express the following relation:

A | MALE ACCESSORY REPRODUCTIVE ORGANS (Produce semen for suspending the sperm) | = | 2 SEMINAL VESICLES | + | 2 BULBOURETHRAL GLANDS | + | 1 PROSTATE GLAND

Organ 2

Tracing the Path of a Sperm Cell

Bringing together the essential facts from the two preceding sections, we can provide a brief flow-diagram tracing the pathway followed by an ejaculated sperm cell (spermatozoon):

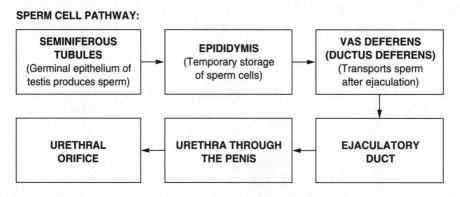

SPERM CELL PATHWAY:

| SEMINIFEROUS TUBULES (Germinal epithelium of testis produces sperm) | → | EPIDIDYMIS (Temporary storage of sperm cells) | → | VAS DEFERENS (DUCTUS DEFERENS) (Transports sperm after ejaculation) |

| URETHRAL ORIFICE | ← | URETHRA THROUGH THE PENIS | ← | EJACULATORY DUCT |

Internal Anatomy & Physiology of the Female Reproductive Pathway

So far, we have proceeded as far as ejaculation of spermatozoa and semen from the external urethral orifice of the penis. Humans engaging in *coitus* (**KOH**-ih-tus) must have their external genital organs "come together" (*coit*) during sexual intercourse. This coitus (sexual intercourse), if successful (reproductively speaking), achieves the *internal fertilization* of an *ovum* (**OH**-vum) – an "egg" (*ov*) cell deep within the female.

The penis of the male must be inserted into the *vagina* (vah-**JEYE**-nah). Reflecting the war-like orientation of the Early Roman scholars, the word, *vagina*, actually means "sheath" (as in a sheath that holds a penis "sword"). The penis-"ensheathing" vagina serves as an entryway for ejaculated spermatozoa that may produce the internal fertilization of an ovum.

Of course, remember the old saying that, "The door swings both ways!" Now, the word, clitoris, actually means "door-tender," and Figure 15.3 showed that it does lie just above the vaginal orifice. Thus, the vaginal orifice (opening) is the "door," isn't it? The sole function of the clitoris is sexual arousal in the female. Like the penis (its sexual homologue), the clitoris has a *spongy body* filled with *venous sinuses*. Its surface is also densely covered with sensory nerve endings. Thus, the clitoris (like the penis) becomes erect and stiff during sexual arousal. ("*Tsk! Tsk!* We can already see your face getting red, Baby Heinie!)

So, when the clitoris gets erect, it (as the "door-tender") moves out of the way of the "door" (vaginal orifice), so that the erect penis may be properly inserted into its "sheath." Figure 15.5 shows the internal anatomy of the female reproductive pathway, and what happens during coitus – when both man and woman "come together."

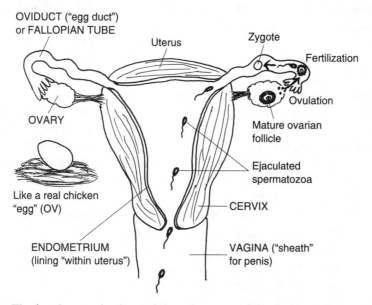

Fig. 15.5   The female reproductive pathway after successful coitus ("coming together").

Internal Fertilization: Mr. Sperm Knocks On Her "Door," and Ms. Ovum Lets Him In!

Figure 15.5 shows some spermatozoa being ejaculated into the vagina, and then passing up through the *cervical* (**SIR**-vih-kal) *os* (**OHS**). A *cervix* (**SIR**-viks), in general, is any narrow, "neck" (*cervic*)-like structure. One of these is the *uterine* (**YOU**-ter-in) *cervix* – the tapered, "neck"-like, inferior portion of the *uterus* (**YOU**-ter-us). The uterus is the technical name for the "womb" (*uter*).

An *os* (**OHS**), in general, is some kind of "mouth" or "opening." Hence, the cervical os is the little mouth-like opening at the base of the uterine cervix, the hole through which ejaculated spermatozoa may travel.

THE OVARIES: THEY'RE BOTH BUSY "DROPPING" TINY "EGGS"

Attached to the top of the uterus on either side are the right and left *oviducts* (**OH**-vih-ducts) – "egg ducts." The oviducts are alternately called the *Fallopian* (fah-**LOH**-pea-un) *tubes* in honor of their discoverer, the Italian anatomist Gabriello Fallopio (fah-**LOH**-pea-oh). The oviducts (Fallopian tubes) are a pair of slender egg ducts that carry released *ova* (**OH**-vah) – "egg" cells – towards the uterus. The source of these ova are the right and left *ovaries* (**OH**-var-**eez**). Each *ovary* (**OH**-var-**ee**) is named for its oval, whitish appearance, much like a chicken "egg" (*ovari*).

About once a month, starting in puberty, the fertile female has a sudden surge in trophic hormone secretion from the anterior pituitary gland (Chapter 10). A particular trophic hormone called *luteinizing* (**LOO**-teh-**neye**-zing) *hormone* or *LH*, circulates to the ovaries. Here, LH dissolves and weakens the wall of an ovary just enough to trigger *ovulation* (**ahv**-you-**LAY**-shun), the process of dropping or releasing a "little egg" (*ovul*) from the ovarian surface.

The released ovum is usually swept up into a nearby oviduct (Fallopian tube). Fertilization generally occurs in the first (outer) 1/3 of the oviduct. Fusion of sperm and ovum together creates a *zygote* (**ZEYE**-goat). The zygote literally means an ovum plus sperm cell nucleus "yoked together" to create a single new cell having 46 chromosomes:

THE ZYGOTE = SPERM CELL + OVUM CELL
(A single new   NUCLEUS   NUCLEUS
cell with 46   (23 chromosomes) (23 chromosomes)
chromosomes)

Cell 1

Development Leading to Birth: The Cute Little "Sweller" Gets Ready to Stick His Head Out!

Thanks to internal fertilization of an ovum, we all start our lives as a single cell, the zygote. But the zygote doesn't remain a single cell for very long! As it moves through the oviduct and towards the uterus, it undergoes a series of *mitoses* (my-**TOH**-seez). The zygote just keeps dividing into two cells, four cells, eight cells, on-and-on. As the name, embryo, literally means a "sweller," the zygote (beginning embryo) keeps dividing and "swelling" in size, in greater Biological Order, and in the growing pattern of its complexity.

EARLY STAGES OF EMBRYO DEVELOPMENT

Figure 15.6 outlines the major developments leading from the zygote to the *fetus* (**FEE**-tus) or final, human-like "offspring."

The single-celled zygote becomes a solid mass of cells called a *morula* (mor-**OO**-lah). The word, morula, translates into Common English to mean "little mulberry." The morula, then, looks somewhat like a real mulberry hanging on a bush, which has many little bumps or "cells" on its fruit surface.

The morula eventually passes out of the oviduct and enters the *body* (main hollow cavity) of the uterus. Here it becomes a *blastula* (**BLAS**-chew-lah), also called a *blastocyst* (**BLAS**-toh-**sist**). The blastula is a "little sprouter" (*blastul*) or "hollow sprouting bladder" (*blastocyst*). To be sure, the blastula, being called a blastocyst, with *cyst* meaning hollow "bladder," looks quite a bit like a hollow raspberry inside, instead of a solid mulberry!

We get the idea of "sprouter" from what the blastula *does*. It *implants* itself into the *endometrium* (**en**-doh-**ME**-tree-um), the "inner" (*endo-*) epithelial lining of the "uterus" (*metr*). After it implants, the blastula sends out spreading *roots* of cells (like a sprouting plant), thereby firmly anchoring it into the endometrium.

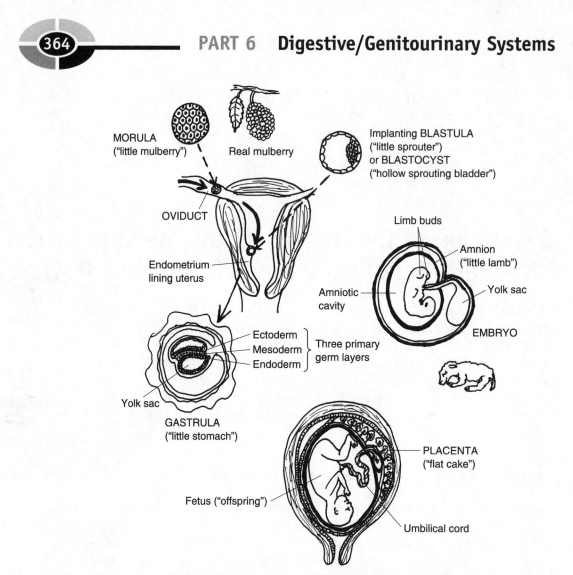

Fig. 15.6 An embryo quickly "swells" into a fetus.

The next stage of embryo development is the *gastrula* (**GAS**-true-lah) or hollow "little stomach" (*gastrul*). As Figure 15.6 clearly displays, the gastrula gives rise to the three *primary germ layers* of the embryo's tiny developing soma (body).

"Hey! I remember the soma and those three primary germ layers, Prof! We talked about them way back in Chapter 9, *The Nervous System and Organs of the 'Special Senses'*!" Yes, Baby Heinie! You have a wonderful memory! And do you recall the *names* of the three primary germ layers in the gastrula? Further, do you remember what general *types* of *adult body tissues* that each of these primary germ layers specialized or differentiated into? [**Study sugges-**

tion: Before peeking again at Figure 15.6, try to answer the two questions just posed by Professor Joe. It is also highly recommended that you flip way back in the book and examine Figure 9.2, which shows, "How the gastrula forms the body."]

Primary germ layer review

To obtain a bite-size review of the gastrula and its three primary germ layers, just recite the simple word equation, below:

A

THE GASTRULA = ENDODERM + MESODERM + ECTODERM
(A "Little Stomach"   (*"Inner skin"*)   (*"Middle skin"*)   (*"Outer skin"*)
with 3 primary germ
  layers in its walls)

Tissue 1

WHAT LIES BEYOND THE GASTRULA?

From our discussion, we can say that the gastrula is basically a three-layered embryo. The gastrula stage usually occurs during the second week after fertilization. Figure 15.7 displays some of the developments that occur thereafter.

A cylinder-shaped body appears

As the developing cell layers in the gastrula divide and widely migrate, the three germ layers roll up into a hollow cylinder, which becomes the *General*

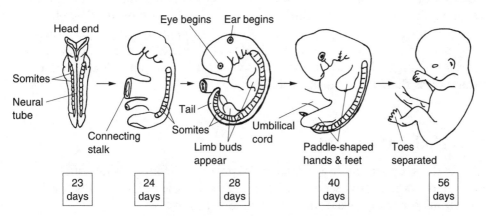

Fig. 15.7   The embryo develops beyond the gastrula.

Cylinder-shaped Body Plan. At this stage, in other words, we humans look much like a chubby little caterpillar – one with a stubby tail!

At just after 3 weeks (around the 23rd day), the cylinder-shaped body contains a *neural tube* in its back, flanked by rows of *somites* (developing body segments).

Membranes appear around the body

At about the same time as the General Cylinder-shaped Body is forming, a number of membranes appear around the Body. A glance back at Figure 15.6, for example, shows the *amnion* (**AM**-nee-un). The amnion is a protective membrane forming a sac around the embryos of "little lambs" (*amnions*), human beings, and other types of *mammals*. The amnion also encloses a quantity of *amniotic* (**am**-nee-**AH**-tik) *fluid*. This fluid creates a watery cushion and shock absorber for the embryo, as well as keeping the body wet and moist.

Another membrane becomes the *yolk sac*, which provides nourishment for the early embryo. By 24 days, a *connecting stalk* appears in the middle of the now worm-like body. The yolk sac hangs off to one side of this connecting stalk. Both attach to a primitive *placenta* (plah-**SEN**-tah), a "flat cake" (*placent*) of highly vascular (blood vessel-rich) tissue that nourishes the developing embryo and later, the fetus.

Organogenesis goes into full swing

During the same time that the membranes (such as the amnion) are forming around the *outside* of the embryo, *organogenesis* (**or**-gan-oh-**JEN**-eh-sis) – "production of" (-*genesis*) living "organs" (*organo-*) – is proceeding at a fast pace *inside* of the embryo. By 24 days ($3\frac{1}{2}$ weeks), for instance, a miniature heart organ has developed and is vigorously pumping blood for a body that is less than 1/4 of an inch long!

Limb buds sprout

By 28 days, upper and lower *limb buds* sprout out from the sides of the grub-like body, which now seems to be getting some eyes! These buds mark the locations for later development of true body limbs.

An umbilical cord appears

By 40 days, an *umbilical* (um-**BILL**-ih-kal) *cord* has replaced the yolk sac and connecting stalk, and has firmly rooted itself into the nourishing placenta. Upper and lower appendages have grown out from the limb buds, and the hands and feet look like little paddles. A definite eye with pigment is now visible.

End of embryo stage: All major organ systems have developed

By 56 days (8 weeks or 2 months), the embryo body has a distinctly human form. The tail is long gone, and the fingers and toes are completely separated from one another in the hands and feet.

"Can my little brother *see* me, yet?" Not quite yet, Baby Heinie. The eyes are present and are well-developed, but the eyelids are stuck together! "You *mean*...They're keeping my little brother in the *dark*?"

After about 56 days (2 months or so), organogenesis has been completed. All major body tissues and organ systems are up and functioning! The embryo stage is nearly *over*! (It officially ends at 9–12 weeks – the third month.)

Good! My Mom is really pregnant! Grandma says she's 8 weeks along! Can I see my little brother's *face*, Good Professor Joe?" Okay, here he is!

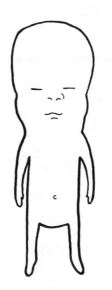

"*Yikes!* Why is my little brother's *head* so big? Is that *normal* gross anatomy, Prof?" At this stage of development, it is normal. At 8 weeks, the head is almost as big as the rest of the body! But, not to worry! The rest of the body will rapidly lengthen.

"Well, how *long* is my little brother at 8 weeks, and how much does he *weigh?*" At the end of the embryo stage, the body is about 1.2 inches (or 30 mm) long, and it weighs about 2 grams (0.06 ounce).

"Well, maybe my little brother can't *see*, but at least he won't be born with a *weight* problem!"

SNAPSHOT SUMMARY OF THE EMBRYO STAGE

Let us "demystify" and simplify the preceding complex stages of embryo development, by means of a descriptive flow chart:

THE SEQUENCE OF STAGES IN EMBRYO DEVELOPMENT:

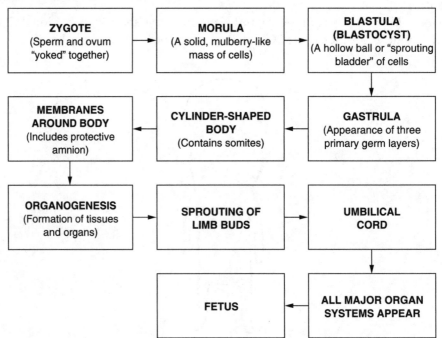

ENTER THE FETUS AND NEONATE

"What happens after the embryo, Prof?" The next stage is the fetus or "off-spring." Basically, what we're dealing with, here, is a miniature version of a human being, who has all organ systems and tissues present. The main activity now is not tissue differentiation, but mainly mitosis and a need for further body growth. And at the end of about 9 months, the typical human *gestation* (jes-**TAY**-shun) or "bearing" *period*, your little brother (Baby Heinie Junior) will finally become a *neonate* (**NEE**-oh-**nayt**).

"A **nee**-ohh *whooosie*?" A neonate – a "newborn" child. And if your Mom is going to name your little brother Baby Heinie *Junior*, them I'm *outa'* here! It's high-time to bring this chapter (and *ANATOMY DEMYSTIFIED*) to a merciful *end*!

Heartbreak Hill: Gross Pathological Anatomy During Development

Since the *embryonic* (**em**-bree-**AH**-nik) *period* is when most tissue differentiation and appearance of organ systems takes place, it is also the most susceptible period during human development for damage from various causes. At the beginning of this book, we talked much about Biological Order and recognizable patterns of body structure. But what if these patterns of Order are disrupted during the embryonic stage? Can we still expect that Geometric Bodyspace will be preserved, undisturbed, within the neonate?

PHOCOMELIA AND THALIDOMIDE

One very dramatic and disturbing example of the effects of a potent *teratogen* (ter-**AT**-oh-jen) or "monster" (*terat*) "producer" (*-gen*) on the developing embryo is provided by the drug called *thalidomide* (thah-**LID**-oh-meyd). Thalidomide was used extensively as a sedative and sleeping pill during the early 1960s. Physicians had no qualms about giving it to women in the early stages of pregnancy! Its use was soon discontinued, however, when a significant number of "thalidomide babies" had been born.

All too often, such "thalidomide babies" suffered the severe birth defect called *phocomelia* (**foh**-koh-**MEE**-lee-ah) or "seal" (*phoc*) "limbs" (*mel*). A neonate with this *congenital* (kahn-**JEN**-ih-tal) *malformation* ("bad formation") had it present "with" (*con-*) "birth" (*genit*). The stricken child was born missing the proximal portions of his limbs, so that his hands and feet were attached directly to the body trunk, giving the appearance of having seal flippers (see Figure 15.8).

Apparently, thalidomide interfered with the normal development and maturation of the mesoderm, especially, which is mainly responsible for healthy bone and muscle tissue differentiation.

Although it is no longer approved for use by pregnant women, the severe congenital deformities caused by thalidomide point to the importance of proper *prenatal* (pree-**NAY**-tal) or "before birth" (*nat*) care.

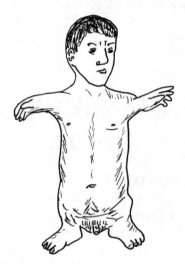

Fig. 15.8 Phocomelia: Stubby "seal limbs" in a stricken child.

Quiz

Refer to the text in this chapter if necessary. A good score is at least 8 correct answers out of these 10 questions. The answers are listed in the back of this book.

1. The term, genitourinary, is often used because:
   (a) Neither the reproductive nor urinary systems can be accurately described
   (b) "It takes one, to know one!"
   (c) Both the genital and urinary systems share many features in common

(d) No other terms are available

2. The three major zones of the kidney:
    (a) Renal cortex, lymphatic nodules, renal medulla
    (b) Adrenal crust, umbilical leads, pyramidal zone
    (c) Renal pelvis, medulla, and cortex
    (d) Hypothalamus, renal hilus, benign nevus

3. Another name for the urinary bladder:
    (a) Urocyst
    (b) Urethra
    (c) Renal capsule
    (d) Cholecyst

4. The common source for development of the external genital organs in both males and females:
    (a) Labia minora
    (b) Prepuce
    (c) Glans penis
    (d) Genital tubercle

5. The spermatozoa are actually produced within the:
    (a) Epididymis
    (b) Seminiferous tubules
    (c) Goblet cells
    (d) Ejaculatory duct

6. Creates a powerful sucking action to draw sperm out of the epididymis:
    (a) Vas deferens
    (b) Ductus defense
    (c) Germinal epithelium
    (d) Ureter

7. Literally translates to mean a "sheath" for the penis:
    (a) Clitoris
    (b) Vagina
    (c) Urethra
    (d) Uterus

8. Represents a sperm and ovum nucleus "yoked" by fertilization:
    (a) Zygote
    (b) Morula

(c) Blastula
(d) Renal cortex

9. The stage of development that embeds in the uterine endometrium:
(a) Gastrula
(b) Cylinder-shaped body
(c) Stage with limb buds
(d) Blastocyst

10. The embryonic stage essentially ends when:
(a) The zygote has stopped dividing
(b) Teratogens naturally reverse the developmental process
(c) All organ systems are present
(d) Three primary germ layers appear

Body-Level Grids for Chapter 15

Several key body facts were tagged with numbered icons in the page margins of this chapter. Write a short summary of each of these key facts into a numbered cell or box within the appropriate *Body-Level Grid* that appears below.

Anatomy and ***Biological Order*** **Fact Grids for Chapter 15:**

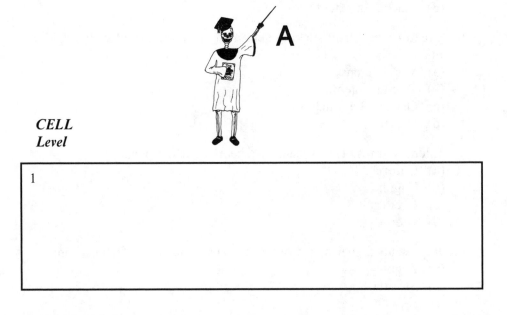

CELL
Level

1

TISSUE
Level

1

ORGAN
Level

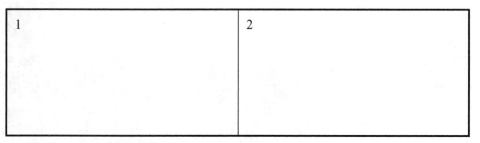

ORGAN SYSTEM
Level

1

Test: Part 6

DO NOT REFER TO THE TEXT WHEN TAKING THIS TEST. A good score is at least 18 (out of 25 questions) correct. Answers are in the back of the book. It's best to have a friend check your score the first time, so you won't memorize the answers if you want to take the test again.

1. The digestive system is also called the:
   - (a) Renal tubules
   - (b) Alimentary canal
   - (c) Globus pallidus
   - (d) Sigmoid colon
   - (e) Respiratory tract

2. The opposite of ingestion:
   - (a) Digestion
   - (b) Absorption
   - (c) Secretion
   - (d) Reabsorption
   - (e) Egestion

3. Movement of material from the lumen of the digestive tract, into the bloodstream:
   (a) Secretion
   (b) Osmotic pressure
   (c) Absorption
   (d) Defecation
   (e) Mastication

4. Chemical digestion of ____ begins in the oral cavity:
   (a) Carbohydrates
   (b) Lipids
   (c) Fats
   (d) Proteins
   (e) Glucose

5. A common passageway shared by both the digestive and respiratory tracts:
   (a) Esophagus
   (b) Cholecyst
   (c) Pharynx
   (d) Epiglottis
   (e) Trachea

6. The submucosa:
   (a) Lines the lumen of the intestinal tract
   (b) Seldom participates in the process of digestion
   (c) Circulates absorbed nutrients throughout the digestive tube wall
   (d) Consists primarily of circular smooth muscle tissue
   (e) Acts like a serous membrane

7. Hangs down from the greater curvature of the stomach:
   (a) Lesser omentum
   (b) Larynx
   (c) Greater omentum
   (d) Cardia
   (e) Gastric folds

8. The gastric juice is rich in:
   (a) HCO_3^-
   (b) HCl
   (c) Salivary amylase
   (d) Fungus
   (e) OH^- ions

9. The "gatekeeper" to the small intestine:
   (a) Pyloric sphincter
   (b) Hepatopancreatic sphincter
   (c) Cecum
   (d) Transverse colon
   (e) Hepatic flexure

10. Superior portion of the stomach just below the entrance of the esophagus:
    (a) Cardia
    (b) Body
    (c) Greater curvature
    (d) Rugae
    (e) Peyer's patches

11. Literally means "presence of 12":
    (a) Duodenum
    (b) Ilium
    (c) Jejunum
    (d) Sigmoid colon
    (e) Rectum

12. Process by which bile helps digest fatty foods:
    (a) Purification
    (b) Facilitated diffusion
    (c) Acid–base balancing
    (d) Emulsification
    (e) Vomiting

13. Bile is actually secreted by the:
    (a) Hypothalamus
    (b) Anterior pituitary
    (c) Pancreatic acinar cells
    (d) Beta cells
    (e) Liver cells

14. The ultimate breakdown products of carbohydrates, that are absorbed:
    (a) Fatty acids
    (b) Chyme
    (c) Glycerol
    (d) Simple sugars
    (e) Steroids

15. Tiny bumps present on "tufts of hair":
    (a) Microvilli
    (b) Tubuloadenomas
    (c) Villi
    (d) Genital tubercles
    (e) Mammillary bodies

16. The middle section of the small intestine:
    (a) Ilium
    (b) Rugae
    (c) Jejunum
    (d) Duodenum
    (e) Ileo-cecal sphincter

17. Chyme is modified into ____ within the colon:
    (a) Chewing gum
    (b) Feces
    (c) Food boluses
    (d) Salivary amylase
    (e) Pepsin

18. A straight tube leading directly to the anus:
    (a) Rectum
    (b) Ascending colon
    (c) Esophagus
    (d) Oral cavity
    (e) Pharynx

19. "Seed animals" ejaculated by males:
    (a) Androgens
    (b) Ova
    (c) Spermatozoa
    (d) Semens
    (e) Spicules

20. The major microscopic functional units of the kidneys:
    (a) Renal pyramids
    (b) Fallopian tubes
    (c) Nephrons
    (d) Renal capillaries
    (e) Mucosae

21. Temporarily stores urine before voiding:
(a) Urocyst
(b) Glomerulus
(c) Renal capsule
(d) Renal pelvis
(e) Medulla oblongata

22. Sexual homologues:
(a) Clitoris and glans penis
(b) Scrotum and anus
(c) Testes and vagina
(d) Ova and ovaries
(e) Tonsils and parathyroids

23. Responsible for sexual arousal in females:
(a) Flowers and candy
(b) Clitoris
(c) Perineum
(d) Internal anal sphincter
(e) Cholecyst

24. Temporarily stores sperm cells between ejaculations:
(a) Ejaculatory duct
(b) Urethra in penis
(c) Epididymis
(d) Seminiferous tubules
(e) Ductus deferens

25. A thick, milky, sugar-rich fluid that nourishes and suspends the spermatozoa:
(a) Semen
(b) Bile
(c) Serous drainage
(d) Vomitus
(e) Renal filtrate

Final Exam

DO NOT REFER TO THE TEXT WHEN TAKING THIS EXAM. A good score is at least 75 correct. Answers are in the back of the book. It's best to have a friend check your score the first time, so you won't memorize the answers if you want to take the test again.

1. An organism in general is a living body with:
   (a) Disordered and uncoordinated organ systems
   (b) Biological Order and recognizable patterns
   (c) Contrasting surges of energy which do not dovetail
   (d) Loss of control over its internal environment
   (e) Most of its levels being non-living

2. The word, anatomy, comes from the Ancient Greek for:
   (a) "The process of cutting (something) up or apart"
   (b) "Anna, go get Tommy!"
   (c) "Neutral substance"
   (d) "All things are for the best"
   (e) "Control of sameness"

3. Frequently given credit as being the Father of Anatomy:
   (a) Claude Bernard
   (b) Harvey Cushing
   (c) Max Planck
   (d) Hippocrates
   (e) Andreas Vesalius

4. One of the basic characteristics of a structure:
   (a) Doesn't occupy any space
   (b) Under the force of gravity, it exhibits no weight
   (c) Its underlying "skeleton" can be modeled as a woven cloth or fabric
   (d) Only occurs under resting conditions
   (e) Must always exist in harmony with other structures

5. The sentence, "The ball flew across the fence," is best cited as an example of:
   (a) Physiology
   (b) Structure
   (c) Subatomic layering
   (d) Function
   (e) Anatomy

6. "The hemoglobin molecule carries O_2," is most appropriately classified as representing:
   (a) Homeostasis
   (b) Function
   (c) Negative feedback control
   (d) Anatomic planing
   (e) Physiology

7. The observation that turtles, like humans, have a vertebral column:
   (a) Biochemistry
   (b) Comparative anatomy
   (c) Vertebrate physiology
   (d) Mammalian anatomy
   (e) Embryology

8. Alternately known as developmental anatomy:
   (a) Histology
   (b) Renology
   (c) Embryology
   (d) Genetics

(e) Geology

9. Stated that, "The Book of Nature is written in characters of Geometry":
    (a) Galileo
    (b) Emilia Lockhart
    (c) Baby Heinie
    (d) Albrecht Dürer
    (e) Thomas Aquinas

10. Level found at the base of the Great Body Pyramid:
    (a) Cell
    (b) Atom
    (c) Organelle
    (d) Molecule
    (e) Subatomic particle

11. Microscopic anatomy typically includes:
    (a) Organs
    (b) Organ systems
    (c) Organisms
    (d) Ecological relationships
    (e) Cells

12. Physiology begins here:
    (a) Organelles
    (b) Taxonomic orders
    (c) Molecules
    (d) Cells
    (e) Viruses

13. A torn gastrocnemius (calf) muscle:
    (a) Biological Order at the organism level
    (b) Biological Disorder at the organ level
    (c) Cosmic Order throughout the universe
    (d) Human physiology gone bad
    (e) Complementarity between body structure and function

14. Mummy cases, hieroglyphics, and anatomic planes:
    (a) Found only in the time of Ancient Egypt
    (b) Largely based upon the concept of a rectangle
    (c) Homeostasis of the internal environment
    (d) All are also anatomic sections
    (e) Require thorough knowledge about birds

15. There is one and only one midsagittal plane, because:
    (a) Transverse planes inevitably result in transverse sections
    (b) The body can only be subdivided equally, left-from-right, along a single vertical line
    (c) There are two sagittal sutures
    (d) Numerous parasagittal planes exist
    (e) Coronal planes aren't as easy to use

16. The tip of the nose is ____ to the forehead:
    (a) Dorsal
    (b) Anterior
    (c) Cephalad
    (d) Interior
    (e) Superior

17. The body trunk = _____:
    (a) Thorax + abdomen + femur
    (b) Skull periphery + gonads + aortic arch
    (c) Abdomen + pelvis + thorax
    (d) Head + neck + lips
    (e) Vertebral column only

18. "Bellybutton lint" would most likely be deposited in this abdomino-pelvic region:
    (a) Left hypochondriac
    (b) Right iliac
    (c) Hypogastric
    (d) Epigastric
    (e) Umbilical

19. Pathological anatomy generally reflects:
    (a) A state of Clinical Health
    (b) Homeostasis and bilateral symmetry
    (c) Broken patterns of body structure
    (d) Intact levels of body organization
    (e) Smooth functioning of the human genes

20. The Chemical Level includes:
    (a) Cells, proteins, and mitochondria
    (b) Sodium ions, electrons, and glucose
    (c) Heart, lungs, and kidneys
    (d) Areolar connective tissue
    (e) Circulatory and respiratory systems

21. NaCl is a well-known:
    (a) Organic molecule
    (b) Body solvent
    (c) Isotonic solution
    (d) Electrolyte
    (e) Carbon-bearer

22. The extracellular matrix basically consists of:
    (a) Dry waste matter
    (b) A complex meshwork of structural proteins outside cells
    (c) The ICF
    (d) Pure H_2O
    (e) Tissue enzymes

23. The lipid family includes the:
    (a) Amino acids
    (b) Enzymes
    (c) Structural proteins
    (d) Collagen fibers in the skin
    (e) Phospholipids in cell membranes

24. The storage form of glucose in liver and muscle cells:
    (a) Cellulose
    (b) Triglyceride droplets
    (c) Glycogen
    (d) Glucagon
    (e) Cholesterol

25. The nucleic acid containing the sugar, ribose:
    (a) DNA
    (b) Lactic acid
    (c) RNA
    (d) Glycerol
    (e) CO_2

26. Proposed the Modern Cell Theory:
    (a) Watson & Crick
    (b) Lewis & Clark
    (c) Robert Hooke
    (d) Schleiden & Schwann
    (e) Aristotle

27. The cytoplasm basically consists of:
    (a) Cytoskeleton plus intracellular fluid

    (b) Organic solutes with no solvent
    (c) Polar substances only
    (d) Disintegrated pieces of the plasma membrane
    (e) Centrioles

28. Major ATP producer in cells with aerobic metabolism:
    (a) Lysosome
    (b) Ribosome
    (c) Nucleus
    (d) Mitochondrion
    (e) Pore

29. The entire life span of a particular cell:
    (a) Mitosis
    (b) Anaphase
    (c) Cell cycle
    (d) Interphase
    (e) Summation

30. A tissue is best characterized as:
    (a) A collection of similar cells plus the intercellular substance between them
    (b) Interstitial fluid plus ions
    (c) Cytoplasm with organelles functioning as coordinated pairs
    (d) Two or more cells with unlike features
    (e) ECF matter

31. The Principle of Cellular Pathology:
    (a) Structure and function are complementary
    (b) Anatomy is nothing without physiology!
    (c) Pathological anatomy at the cellular level often leads to disease
    (d) Lower levels of body organization have no influence upon higher levels
    (e) Pathophysiology starts to show up at the organelle level

32. A collection of related organs, which interact together to carry out some complex body function:
    (a) Tissue
    (b) Inclusion
    (c) Artifact
    (d) Organ system
    (e) Molecule

33. Provides our main protection from the sun:
    (a) Lignin
    (b) Second messenger
    (c) Melanin
    (d) Keratin
    (e) Glycogen

34. Main fibrous connective tissue portion of the skin:
    (a) Dermis
    (b) Subdermis
    (c) Epidermis
    (d) Colloid
    (e) Adipocyte layer

35. Portion of the endoskeleton that lies along the body's central longitudinal axis:
    (a) Appendages
    (b) Exoskeleton
    (c) Carapace
    (d) Axial skeleton
    (e) Appendicular skeleton

36. Bones of the wrist:
    (a) Carpals
    (b) Tarsals
    (c) Hyoid
    (d) Pelvic
    (e) Cranial

37. Mainly found within the medullary cavity of long bones:
    (a) Dense bone tissue
    (b) Yellow marrow
    (c) Periosteum
    (d) Muscularis
    (e) Tunica adventitia

38. Fibrous joints are also known as:
    (a) Diarthroses
    (b) Freely movable
    (c) Synarthroses
    (d) Partially movable
    (e) Phalanges

39. The biceps femoris is named for the criteria of:
    (a) Body location and action
    (b) Number of heads and direction of fibers
    (c) Points of attachment and relative size
    (d) Number of heads and body location
    (e) Association with mythical characters and geometric shape

40. During contraction, the ____ end of a muscle moves towards the ____ end:
    (a) Origin; body
    (b) Insertion; flexor
    (c) Extension; adduction
    (d) Insertion; origin
    (e) Fascicle; joint

41. Works with a particular prime mover:
    (a) Antagonist
    (b) Synergist
    (c) Agonist
    (d) Levator
    (e) Cooperon

42. The main mass of the body, consisting primarily of bones, skeletal muscles, skin, and joints:
    (a) Viscera
    (b) Gross soma
    (c) Articulations
    (d) Abdominopelvic quadrants
    (e) Serosa

43. The embryonic germ layer that differentiates to form tissue lining the digestive tract:
    (a) Ectoderm
    (b) Mesoderm
    (c) Echinoderm
    (d) Cuticle
    (e) Endoderm

44. Importantly involved in carrying out the response to stress and "Fight-or-Flight":
    (a) Parasympathetic nerves
    (b) Salivary glands
    (c) Sympathetic nerves

    (d) Pancreas
    (e) Parathyroids

45. The auditory nerve:
    (a) Movement of the face
    (b) Sense of vision
    (c) Pain sense
    (d) Sense of hearing
    (e) Movement detection

46. Creates the top of the brainstem:
    (a) Thalamus
    (b) Pons
    (c) Olfactory nerve
    (d) Hypothalamus
    (e) Cerebellum

47. The "tough mother" forming a protective membrane around the CNS:
    (a) Pia mater
    (b) Subarachnoid space
    (c) Cauda equina
    (d) Arachnoid mater
    (e) Dura mater

48. Endocrine gland portion of the pancreas:
    (a) Islets of Langerhans
    (b) Acini
    (c) Globular proteins
    (d) Hematocrit
    (e) Thymic

49. A deficient or below normal secretion:
    (a) Glandular hypertrophy
    (b) Hyposecretion
    (c) Normosecretion
    (d) Hypersecretion
    (e) Adenoma formation

50. Stores and releases antidiuretic hormone into the general bloodstream:
    (a) Posterior pituitary
    (b) Hypothalamus
    (c) Anterior pituitary

    (d) Adrenal cortex
    (e) Adrenal medulla

51. The source for insulin secretion:
    (a) Pancreatic acinar cells
    (b) Thyroid
    (c) Testes
    (d) Alpha cells
    (e) Beta cells

52. Greatly raises the Basal Metabolic Rate:
    (a) ADH
    (b) Thyroxine
    (c) Aldosterone
    (d) Estrogen
    (e) Testosterone

53. Source of Releasing Hormones (RHs):
    (a) Thalamus
    (b) Adrenal medulla
    (c) Adrenal cortex
    (d) Hypothalamus
    (e) Anterior pituitary

54. Gland ablation means gland:
    (a) Removal
    (b) Stimulation
    (c) Fusion
    (d) Development
    (e) Replacement

55. The circulatory system includes the:
    (a) Thymus
    (b) Vascular network
    (c) Vermiform appendix
    (d) Pineal body
    (e) Third ventricle

56. The heart is located within the:
    (a) Lateral ventricle
    (b) Mediastinum
    (c) Pons
    (d) Renal hilus
    (e) Sympathetic ganglia

57. Alternately called the visceral pericardium:
    (a) Myocardium
    (b) Epicardium
    (c) Endocardium
    (d) Heart valves
    (e) Atrium

58. The thick coat of smooth muscle within a vessel wall:
    (a) Muscularis
    (b) Tunica adventitia
    (c) Lumen
    (d) Tunica media
    (e) Internal elastic membrane

59. Arteries always carry blood:
    (a) Towards the heart
    (b) Through the A-V valves
    (c) Around the apex of the heart
    (d) Away from the heart
    (e) That is bluish-red in color

60. When many arterioles vasoconstrict, they dramatically increase the:
    (a) EKG
    (b) CO
    (c) TPR
    (d) BVD
    (e) Vessel diameters

61. Act like blood storage depots:
    (a) Large veins
    (b) Muscular arteries
    (c) Lymphatic capillaries
    (d) Small venules
    (e) Constricted arterioles

62. A one-way valve named for its three cusps:
    (a) Bicuspid
    (b) Aortic
    (c) R semilunar
    (d) Pulmonary
    (e) Tricuspid

63. Tiny, thin-walled air sacs involved in gas exchange:
    (a) Atria
    (b) Pulmonary alveoli
    (c) Systemic arterioles
    (d) Pulmonary veins
    (e) Glanduli

64. The systemic circulation begins with the:
    (a) Pulmonary valve
    (b) R atrium
    (c) L ventricle
    (d) Myocardium
    (e) L atrium and auricle

65. The special circulation that supplies the myocardium:
    (a) Visceral
    (b) Renal
    (c) Coronary
    (d) Hepatic
    (e) Gastric

66. Crushing heart pain:
    (a) Angina pectoris
    (b) Peripheral edema
    (c) Osteoclast hyperactivity
    (d) Femoral thrombosis
    (e) Popliteal contusions

67. The fluid intercellular material of the blood:
    (a) Hematocrit
    (b) Wandering macrophages
    (c) Plasma
    (d) Thrombi
    (e) Coagulants

68. The so-called buffy coat in a test tube of centrifuged blood:
    (a) RBCs
    (b) Neutrophils only
    (c) Thrombocytes
    (d) Leukocytes
    (e) Globulins

69. The albumins help:
    (a) Blood to clot

(b) Affect net osmosis of water into or out of the bloodstream
(c) Activity of the immune system
(d) Promote tissue autolysis
(e) Discourage cancerous predators

70. A deficiency of platelets would likely result in:
    (a) Greater blood clotting
    (b) Cell mitosis
    (c) Lymph formation
    (d) Renal drainage
    (e) A tendency to hemorrhage

71. Hematocrit represents:
    (a) A certain percentage of WBCs in blood
    (b) The ability of the blood to ward off diseases
    (c) The percent of the total blood volume made up of erythrocytes
    (d) The dry weight of our bones
    (e) The tendency of plasma to clot

72. Named for its resemblance to "clear spring water":
    (a) Serum
    (b) Lymph
    (c) Glomerulus
    (d) Joint fluid
    (e) Fibrous exudate

73. A condition wherein the body has achieved protection from disease:
    (a) Cancer
    (b) Vitality
    (c) Courage
    (d) Homeostasis
    (e) Immunity

74. A "little network" of lymphatic vessels lined by endothelial cells:
    (a) A-V venules
    (b) Reticuloendothelial system
    (c) Hepatopancreatic sphincter
    (d) Rete testis
    (e) Uterine endometrium

75. Gland that reaches its maximum size at puberty, then shrinks by adulthood:
    (a) Thyroid
    (b) Adrenal medulla

(c) Thymus

(d) Pancreas

(e) Bone marrow

76. "Little almonds" of lymphatic tissue located in the back of the throat:
    (a) Spleens
    (b) Tonsils
    (c) Parathyroids
    (d) Nephrons
    (e) Sulci

77. Literally "pertains to breathing again":
    (a) Tumorous
    (b) Ventlatic
    (c) Livid
    (d) Diaphoretic
    (e) Respiratory

78. The lower respiratory tract is frequently nicknamed the:
    (a) "Golden Triangle"
    (b) "Three Mighty Rivers"
    (c) "Point of No Return"
    (d) "Land Down Under"
    (e) "Respiratory Tree"

79. The tongue-shaped opening between the vocal cords:
    (a) Epiglottis
    (b) Larynx
    (c) Cartilage horseshoes
    (d) Laryngeal prominence
    (e) Glottis

80. The main windpipe serving both lungs:
    (a) Primary bronchus
    (b) Pharynx
    (c) Trachea
    (d) Smaller bronchi
    (e) Pulmonary alveoli

81. About 2/3 of ____ (a particular organ) lies to the left of the body midline:
    (a) R lung
    (b) Urocyst
    (c) Heart

      (d) Lips

      (e) Sternum

82. Most bronchioles differ from bronchi in that they lack ____ in their walls:

      (a) Smooth muscle

      (b) Cartilage partial-rings

      (c) Bony outgrowths

      (d) Abnormal epithelial hyperplasia

      (e) Mucus

83. A cluster of tiny air sacs in the lungs:

      (a) Alveolus

      (b) Ectopic focus

      (c) Alveolar sac

      (d) Lobe

      (e) Section

84. The tiniest of all air passageways in the lungs:

      (a) Alveolar ducts

      (b) Respiratory bronchioles

      (c) Lymphatic venules

      (d) Terminal bronchioles

      (e) Pulmonary lobules

85. The only place in the respiratory system where gas exchange occurs:

      (a) Walls of all bronchi

      (b) Respiratory canals

      (c) Internal elastic membrane

      (d) Alveolar–capillary interface

      (e) Ventilatory chambers

86. The bulk flow of air into and out of the lungs:

      (a) Simple diffusion

      (b) Pulmonary ventilation

      (c) Inspiration

      (d) Expiration

      (e) A nice long vacation!

87. ____ produces a "chokehold" on the airways:

      (a) Bronchoconstriction

      (b) Alveolar distention

      (c) Pulmonary insufflation

(d) Bronchodilation
(e) Valvular turgor

88. Parasympathetic nerve activity generally has this effect upon air flow to the alveoli:
    (a) Decreased
    (b) No change
    (c) Enhanced
    (d) Modified to allow for external secretions
    (e) Completely blocked

89. Goes in an opposite direction of secretion into the digestive tube:
    (a) Absorption
    (b) Digestion
    (c) Ingestion
    (d) Egestion
    (e) Defecation

90. *E. coli* released into the abdominal cavity after appendix rupture demonstrates that:
    (a) There is no safe place to live, anymore!
    (b) Serous drainage is no fun!
    (c) In a structural sense, the alimentary tract lies outside of the body
    (d) Two wrongs don't make a right!
    (e) Bacteria are never helpful in the human body!

91. A soft "ball" of ingested food:
    (a) Alveolus
    (b) Chyme
    (c) Bolus
    (d) Dentate
    (e) Villus

92. Responsible for exciting the circular smooth muscle in the bowel wall to contract:
    (a) Cartilage caps
    (b) Longitudinal smooth muscle layer
    (c) Radially attached striated fibers
    (d) Embedded goblet cells
    (e) Forests of waving cilia

93. Stores and releases bile:
    (a) Cholecyst
    (b) Jejunum

   (c) Hepatic cell

   (d) Areolar connective strap

   (e) Hypothalamic secretory neuron

94. Tiny bumps on bigger bumps that increase surface area for absorption:
   (a) Rugae
   (b) Villi
   (c) Labioscrotal folds
   (d) Microvilli
   (e) Tonsils

95. A spongy "tail" that both delivers sperm and excretes urine:
   (a) Clitoris
   (b) Penis
   (c) Glans
   (d) Nephron
   (e) Vulva

96. A hollow pouch that stores urine between "going to the bathroom":
   (a) Cholecyst
   (b) Yolk sac
   (c) Amnion
   (d) Vagina
   (e) Urocyst

97. Sexual homologues:
   (a) Glans penis and clitoris
   (b) Seminal vesicles and swim bladders
   (c) Ureter and urethra
   (d) Umbilical cord and placenta
   (e) Scrotum and anus

98. The labioscrotal swelling in the early embryo differentiates to become the:
   (a) Urethra within penis
   (b) Labia majora
   (c) Labia minora
   (d) Vagina
   (e) Genital tubercle

99. Bulbourethral glands and the prostate are examples of:
   (a) Secondary sex characteristics
   (b) Seminal vesicles

(c) Accessory reproductive organs
(d) External genitals
(e) Umbilical cord uniters

100. The very first stage of a human embryo:
(a) Morula
(b) Blastula
(c) Zygote
(d) Endometrium
(e) Amnion

Answers to Quiz, Test, and Exam Questions

Chapter 1

1. B	2. D	3. A	4. D
5. C	6. A	7. C	8. C
9. B	10. A		

Chapter 2

1. B	2. A	3. D	4. A
5. B	6. D	7. D	8. C
9. D	10. B		

Chapter 3

1. B	2. D	3. D	4. A
5. B	6. A	7. C	8. B
9. B	10. D		

Test: Part 1

1. C	2. B	3. D	4. A
5. B	6. D	7. A	8. B
9. A	10. C	11. C	12. E
13. C	14. D	15. A	16. A
17. C	18. D	19. A	20. D
21. B	22. E	23. C	24. B
25. B			

Chapter 4

1. C	2. D	3. B	4. D
5. B	6. D	7. D	8. B
9. A	10. A		

Chapter 5

1. B	2. C	3. A	4. D
5. C	6. B	7. D	8. C
9. D	10. C		

Test: Part 2

1. D	2. A	3. C	4. B
5. A	6. D	7. B	8. C
9. B	10. B	11. B	12. A
13. D	14. B	15. A	16. C
17. A	18. C	19. B	20. E
21. A	22. C	23. A	24. E
25. D			

Chapter 6

1. C	2. B	3. A	4. D
5. C	6. C	7. A	8. B
9. C	10. D		

Chapter 7

1. D	2. C	3. B	4. D
5. A	6. B	7. D	8. A
9. C	10. D		

Chapter 8

1. D	2. D	3. A	4. B
5. D	6. D	7. A	8. D
9. A	10. A		

Test: Part 3

1. C	2. B	3. A	4. C
5. A	6. E	7. E	8. D
9. D	10. C	11. D	12. E
13. A	14. A	15. A	16. B
17. A	18. B	19. A	20. B
21. D	22. C	23. E	24. E
25. D			

Chapter 9

1. D	2. D	3. B	4. A
5. C	6. A	7. C	8. B
9. C	10. C		

Chapter 10

1. C	2. A	3. C	4. C
5. A	6. B	7. D	8. A
9. D	10. A		

Test: Part 4

1. A	2. E	3. C	4. C
5. D	6. B	7. B	8. A
9. D	10. E	11. A	12. E
13. D	14. C	15. A	16. B
17. A	18. E	19. C	20. B
21. D	22. E	23. B	24. A
25. D			

Chapter 11

1. C	2. D	3. A	4. B
5. A	6. B	7. C	8. D
9. D	10. A		

Chapter 12

1. B	2. D	3. A	4. B
5. A	6. D	7. C	8. A
9. A	10. C		

Chapter 13

1. B	2. D	3. D	4. B
5. C	6. D	7. A	8. C
9. D	10. A		

Test: Part 5

1. A	2. C	3. B	4. C
5. B	6. B	7. A	8. B
9. B	10. C	11. A	12. B
13. A	14. D	15. B	16. A
17. B	18. A	19. D	20. C
21. D	22. D	23. A	24. D
25. E			

Chapter 14

1. C	2. B	3. D	4. B
5. A	6. C	7. C	8. C
9. B	10. A		

Chapter 15

1. C	2. C	3. A	4. D
5. B	6. A	7. B	8. A
9. D	10. C		

Test: Part 6

1. B	2. E	3. C	4. A
5. C	6. C	7. C	8. B
9. A	10. A	11. A	12. D
13. E	14. D	15. A	16. C
17. B	18. A	19. C	20. C
21. A	22. A	23. B	24. C
25. A			

Final Exam

1. B	2. A	3. E	4. C
5. D	6. B	7. B	8. C
9. A	10. E	11. E	12. D
13. B	14. B	15. B	16. B
17. C	18. E	19. C	20. B
21. D	22. B	23. E	24. C
25. C	26. D	27. A	28. D
29. C	30. A	31. C	32. D
33. C	34. A	35. D	36. A
37. B	38. C	39. D	40. D
41. C	42. B	43. E	44. C
45. D	46. A	47. E	48. A
49. B	50. A	51. E	52. B
53. D	54. A	55. B	56. B
57. B	58. D	59. D	60. C

61. A	62. E	63. D	64. E
65. C	66. A	67. C	68. D
69. B	70. E	71. C	72. B
73. E	74. B	75. C	76. B
77. E	78. E	79. E	80. C
81. C	82. B	83. C	84. A
85. D	86. B	87. A	88. A
89. A	90. C	91. C	92. B
93. A	94. D	95. B	96. E
97. A	98. B	99. C	100. C

INDEX